Quantum Physics of Time Travel

Relativity, Space Time, Black Holes, Worm Holes, Retro-Causality, Paradoxes

Joseph Gabriel, Ph.D.

Quantum Physics of Time Travel

Relativity, Space Time, Black Holes, Worm Holes, Retro-Causality, Paradoxes

Joseph Gabriel, Ph.D.

Cosmology Science Publishers, Cambridge
Lan Tao, Managing Editor

Copyright © 2014, 2017
R. Joseph, Ph.D.
Published by: Cosmology Science Publishers, Cambridge, MA

All rights reserved. This book is protected by copyright. No part of this book may be reproduced in any form or by any means, including photocopying, or utilized in any information storage and retrieval system without permission of the copyright owner.

The publisher has sought to obtain permission from the copyright owners of all materials reproduced. If any copyright owner has been overlooked please contact: Cosmology Science Publishers at Editor@Cosmology.com, so that permission can be formally obtained.

ISBN-10: 1938024222
ISBN-13: 9781938024221

Relativity, Space Time...

Contents

1: The Time Machine of Past, Present and Future 8
H.G. Wells: The Time Traveler 8
The Nature Of Time and the Conundrums Of Time Travel 8

2: Time Is Relative: Future, Past, Present Exist Simultaneously 15
The Physics of Time 15

3: Time Dilation And Contraction of Space Time 21
The Future and the Past Are Relative 24
Space-Time Contraction 24
The Contraction of Time 26

4: Twins, Time Travel, Gravity And Aging 31
Inertial Frames 33
The Twins and The Past 35

5: Time Travel And Aging: Clocks, Gravity, Altitude, Longitude & Longevity 37
Gravity Kills 38
Longevity and Latitude 39

6: Acceleration, Light Speed, Time Travel, G-Forces And Warp Drives 40
Time in Time-Space Travel 40
Gravity and Health in Space Travel 41
Fuel and G-Forces 44
Warp Drives 45

7: The Curvature of Space-Time: Gravity and Bending of Light and Time   46
The Circle of Time   48
Gravity Holes in Space-Time   49
Pockets, Warps, And Holes in Space Time   50
Gravity Bends Light   55
Galactic Lensing   59

8: The Circle of Time: In A Rotating Universe The Future Leads To The Past   62
Closed Time Curves In A Rotating Universe: The Future Leads to the Past   64
Multiple Earth's In Curved Space Time   65
Closed Time Light Curves: The Future Causes the Past   65
Patterns, Symmetry, Rotating Universes and the Circle of Time   68
Black Holes, Red Shifts, and Cosmic Evidence For a Rotating Universe   73
Time is a Circle   77

9: Time Travel Through Black Holes in the Fabric of Space-Time   78
Tunneling Through Time   85
Vacuums and Negative Energy   91

10: Microscopic Time Travel At Light Speed   94
Shrinking In Time   94
Length Contraction   94
Photons and the Paradox of Infinite Mass at Light Speed   97
Planck Length and Microscopic Holes in Space-Time   98
Negative Mass and Negative Energy and Duality of Coming and Going   100

11: "Worm Holes" In Extreme Curvatures of Space Time   102
Gravity and the Layered Folding of Space   103
"Worm Holes" In Space-Time   108
Worm Holes and Negative Energy in Space-Time   110
Time Machines Through the Tunnel of Time   113

12: Worm Holes, Negative Energy, Casimir Force And The Einstein-Rosen Bridge   114
How Holes Form Between Layers in Extreme Curvatures of Space-Time   117
The Casimir Vacuum   121
Gravity, Quantum Inequalities and Negative Energy   124
Negative Energy and Transmembrane Time Machines   126
The Negative Energy Negative Mass Time Traveler   128

13: Black Holes & Gravitational Sling Shots   131
The Future Leads to the Past   132

14: The Time Traveler in Miniature: Negative Mass and Energy   139

15: Tachyons, Negative Energy, The Circle of Time: From the Future to the Past   143
Space-Like and Time-Like Separations   144
Tachyons   145
Duality   148
Electrons, Positrons, Tachyons and the Circle of Time   149
The Positive and Negative Time Traveler   152

16: Duality: The Past And Future In Parallel   153
Duality and Multiplicity   154

17: The Mirror of Time: Red Shifts, Blue Shifts and Duality   157

18: Into the Past: Duality, Anti-Matter and Conservation of Energy   161
The Conservation of Mass   164
The Anti-Matter Time Traveler   166
The Laws Of Conservation Are Not Violated By Traveling Into The Past   168
A Negatively Charged Time Traveler May Be Unable to Land in the Past   169

19: Quantum Entanglement And Causality: The Future Effects the Past   170
Is The Future Determined? Can The Past Be Changed?   172
The Laws of Consistency And Uncertainty   173
Quantum Entanglement   174
World Lines, Causality, and Entanglement   175
World Lines of Time, Causality, Entanglement   178
Quantum Entanglement, Gravity and Superluminal Time Travel   181
Gravity, Time, And Entanglement.   183

20: Light, Wave Functions and the Uncertainty Principle: Changing the Future and the Past   185
Cause, Effects and The Uncertainty Principle   185
Time and Quantum Physics   192
Light, Wave Functions, Electrons, Positrons and Entanglement   193
Probabilities and The Wave Function of the Time Traveler   197
Wave Functions   198

21: Paradoxes of Time Travel and the Multiple Worlds of Quantum Physics   201
Everett's Many Worlds   202
Changing the Past: Paradoxes and the Principle of Consistency   205
The Principle of Self-Consistency   207
Paradoxes and Many Worlds 208
The "Many Worlds" Resolution of The Grandmother Paradox   209
Multiple Paradoxes. Effects Negating Causes   209
Information Exists Before it is Discovered   210
Time Travel Through Many Worlds   211
And Yet Another Paradox: The Mirror World of the Past   212

22: Epilogue: A Journey Though The Many Worlds of Time   213

REFERENCES   220

1: The Time Machine of Past, Present and Future

H.G. Wells: The Time Traveler

"It was at ten o'clock to-day that the first of all Time Machines began its career. I gave it a last tap, tried all the screws again, put one more drop of oil on the quartz rod, and sat myself in the saddle. I suppose a suicide who holds a pistol to his skull feels much the same wonder at what will come next as I felt then. I took the starting lever in one hand and the stopping one in the other, pressed the first, and almost immediately the second. I seemed to reel; I felt a nightmare sensation of falling; and, looking round, I saw the laboratory exactly as before. Had anything happened? For a moment I suspected that my intellect had tricked me. Then I noted the clock. A moment before, as it seemed, it had stood at a minute or so past ten; now it was nearly half-past three!"

"I drew a breath, set my teeth, gripped the starting lever with both hands, and went off with a thud. The laboratory got hazy and went dark. Mrs. Watchett came in and walked, apparently without seeing me, towards the garden door. I suppose it took her a minute or so to traverse the place, but to me she seemed to shoot across the room like a rocket. I pressed the lever over to its extreme position. The night came like the turning out of a lamp, and in another moment came to-morrow. The laboratory grew faint and hazy, then fainter and ever fainter. To-morrow night came black, then day again, night again, day again, faster and faster still....Then, in the intermittent darknesses, I saw the moon spinning swiftly through her quarters from new to full, and had a faint glimpse of the circling stars...."

"And so my mind came round to the business of stopping...The peculiar risk lay in the possibility of my finding some substance in the space which I, or the machine, occupied... jamming of myself, molecule by molecule, into whatever lay in my way" (H.G. Wells, "The Time Machine").

The Nature Of Time and the Conundrums Of Time Travel

If a time traveler, sitting in his time machine on the surface of Earth, were to journey just one day into the past or the future, and unless equipped with a space suit, space capsule, and life support, he would die in just a few minutes, irradiated and gasping for breath, alone and abandoned in the wilds of outer space.

Earth, and our solar system are in motion, with Earth having a solar orbital speed of 108,000 km/h (~70,000 mph) and our solar system a speed of 720,000 km/h (450,000 mph) as it circles the Milky Way galaxy. If a time traveler stepped into a time machine in Los Angeles, London, or Beijing, and then set out for the future, he would find himself alone in space millions of miles in front of this planet as it orbits through the cosmos. If absent protective life-support or a space capsule, the time traveler would quickly be reduced to a lifeless corpse.

Even if one were to survive safely ensconced inside a time-space ship, there is the prospect of premature aging; the need to overcome life-threatening g-forces; the "Rip Van Winkle Effect;" miniaturization and the transformation to a state of negative energy and negative mass upon traveling at superluminal speeds into the past; and the alteration of every moment of space time as one journeys through it such that one may travel to "a" past or "a" future" but not "the" past or "the" future.

For example, as dictated by quantum mechanics and the "Copenhagen" and "Many Worlds" interpretations of quantum physics (Bohr, 1934, 1963; De-witt 1971; Everett 1956, 1957; Heisenberg 1958), if a time traveler journeys 100 years to the future he/she will come in contact with and affect and alter the quantum composition of every moment of space-time leading to that future such that the future becomes a different future by his passage to it. If that future is not altered by his journey to it, then that future must have already existed and his journey to it must have occurred in that future before he journeyed to it. The basic tenants of quantum mechanics and the "Copenhagen" and "Many Worlds" interpretations, indicate there may be multiple futures, those which may be altered and those which are not; an assemble of infinite futures which may or may not be altered by events leading to those futures.

Conversely, if a time traveler living in the year 2050 journeys 100 years into the past to 1950, each past (local) moment of that entire 100 years of space-time will become altered by the passage through it. "The" past becomes "a" past shaped by the passage of the time traveler who comes in contact with every moment leading to that past. If that past and every moment leading backwards in time are not altered, then this is because he always journeyed to that past, which is part of the past record. Therefore he must have journeyed to the past before he journeyed to it.

If a time traveler journeys 100 years into the long ago and then decides to stay there, then the past becomes the "present." As that "present" from the moment of his arrival in 1950 progresses forwards 100 years, day by day year by year, to the future date and time when the time traveler was born, and then continuing to the date he left on his journey in 2050, he will leave again for the past which will, for him and all those living at that time, will be the "present." And as 100 years from his day of arrival in 1950 go by, day by day, year by year, leading to the day of his birth and that same future time traveling date in 2050, he will leave again for the past, and then again and again and again such that an infinite number of time travelers might arrive simultaneously in "a" or "the" past or separately in multiple altered pasts which are infinite in number.

Time is relative to the observer (Einstein 1905a,b,c, 1906, 1961). Since there are innumerable observers, there is no universal "past, present, future" which are infinite in number and all of which are in motion.

There is more than one "present" and this is because time is not the same everywhere for everyone, and differs depending on gravity, acceleration, frames of reference, relative to the observer (Einstein 1907, 1910, 1961). Time is relative and there is no universal past. No universal future. And no universal now. The "past" in another galaxy overlaps with the "present" on Earth. The "present" in another galaxy will not be experienced on Earth until the future.

Time is like the weather. The weather is different in Beijing versus Berlin, and so is time. The greater the distance between two locations, the greater the differences in time, and this is because time is linked to locations in space and time-space is in motion (Einstein 1961).

According to Einstein's theorems of relativity (Einstein 1905a,b,c, 1907, 1910, 1961), the past, present and future overlap and exist simultaneously but in different distant locations in the dimension known as space-time, and as such *"The distinction between past, present and future is only an illusion"* (Einstein 1955). Quantum physics, the Uncertainty Principle, the "Many Worlds" interpretation of quantum physics, and what Einstein (1930) called "spooky action at a distance" all call into question the causal distinctions between past, present and future.

Time is perceived. Time is experienced. Even eye-witnesses to a crime can't agree on what they saw, heard, or experienced (Haber & Haber, 2000; Megreya & Burton 2008); though the can all agree that they saw, heard, or experienced something, and the same is true of time. Time is "something," it exists, and therefore it must have energy and a wave function which is entangled with motion, movement, the observer, and the quantum continuum which encompasses space-time.

Time has a fluid consistency. Innumerable futures and pasts exist simultaneously albeit in different locations within space-time all of which are in motion. Observers located in New York, Shanghai, Tokyo, Paris, Mexico City, and on other planets in distant galaxies, are also in motion, as planets spin and orbit the sun, the sun orbits the galaxy, and galaxies move about in the universe. Observers, regardless of what planet, solar system, or in what galaxy they reside, are continually moving though space-time and are continually coming into contact with different times. Observers, planets, and galaxies, like time, also exist in space-time, albeit in different distant locations, and all is relative.

Earth orbits around the sun in a curve. This solar system has a curvature and its motion follows a curving path as it orbits this galaxy. Likewise, space-time is curved and light and time follow that curvature (Einstein 1915a,b, 1961;

Relativity, Space Time...

Gödel 1949a,b). All is in motion and has velocity, but because of this curvature, one may travel in a circle and arrive where they began; and the same is true of time. Time is a circle (Gödel 1949a,b). The past leads to the future and the future can lead to the past.

Space-time, because of the different mass and gravities of distant planets and galaxies, is not just curved, but warped. Gravity and space-time are linked to motion. Therefore, increases in velocity and gravity can not only curve but shrink space-time (Einstein 1905c, 1915a,b; Einstein et al. 1923; Lorentz 1892). Just as two chairs sitting on opposite ends of a carpet can be brought closer together if the carpet is scrunched together and folds up and over itself, distant planets and galaxies, and the present and the future, can be brought closer together by the gravity, acceleration, and the curvature, shrinkage, and folding up of space-time (Einstein 1961).

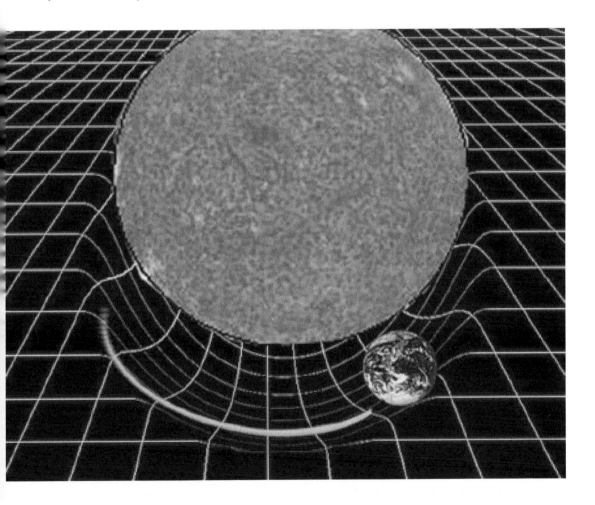

Quantum Physics of Time Travel

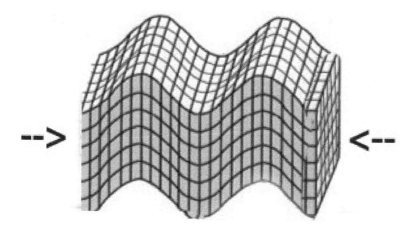

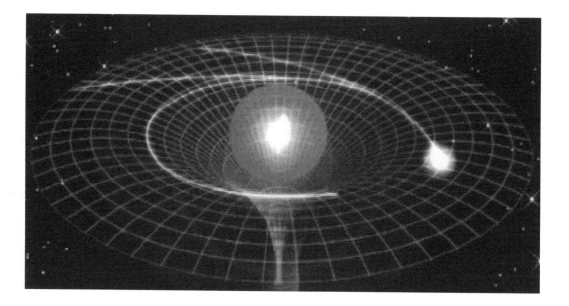

If mass and gravity are sufficiently powerful, space-time doesn't just warp, but may rip, and holes may be punched into the fabric of space-time (Parker & Toms 2009; Ohanian & Ruffini 2013; Thorne & Hawking 1995). These rips and holes which are created by immense gravity and extreme curvatures create passageways which tunnel between these folded up space-time layers. Theoretically (Thorne & Hawking 1995), if a Time-Traveler entered and journeyed through these holes, he may quickly voyage from one distant location to another, from one galaxy to another, and from the present to the future, or to the past or even an parallel universe at velocities faster than the speed of light.

Any time traveler who hopes to visit the future must accelerate toward but remain below light speed. As velocity increases, the distance between the present

Relativity, Space Time...

and the future contracts, and the future arrives more quickly, relative to those left behind. However, those who seek the past must break the cosmic speed limit, for it is only at velocities in excess of light speed that time flows backward and with it, consciousness and memory. And like memory, the past decays.

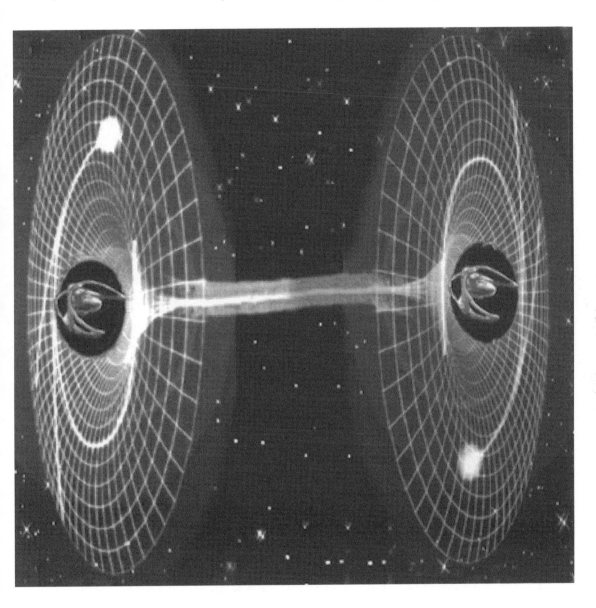

Time is relative to the observer and the experience of the past, future and the "present" are shaped and affected by distance, gravity, acceleration, consciousness, our mood, our surroundings, speeding up under conditions of pleasure and slowing down and sometimes splitting apart or even running backwards under conditions of fear and terror (Joseph 1996, 2010a). Acceleration contracts space-time causing it to speed up or slow down, depending on the location and frame of reference of the observers (Einstein 1905c, 1961; Einstein et al. 1923; Lorentz 1892, 1905). Consciousness, too, can accelerate, particularly under conditions terror, in which case, time slows down and there may be a splitting of consciousness (Joseph 1996, 2010a). Although those with no understanding of consciousness, biology or neurophysiology, might dismiss these alterations of time as "mental distortions," relativity and quantum physics demonstrates otherwise.

"Put your hand on a hot stove for a minute, and it seems like an hour. Sit with a pretty girl for an hour, and it seems like a minute. THAT'S relativity."
-Einstein

The "future" and the "past" are shaped and affected by consciousness which can effect events just by observing them; as illustrated by "entanglement" and Heisenberg's well established Uncertainty Principle (Heisenberg 1927). Consciousness is entangled with the quantum continuum.

Consciousness of the "present" could also be likened to an "event horizon" with the future flowing toward it at an accelerated rate approaching light speed, and the past continuing beyond it at velocities increasingly above the speed of light. The "event horizon" of consciousness, like the "event horizon" of a supermassive black hole, is the "eternal now;" consciousness at the speed of light.

And then there is dream-time where past, present, future and the three dimensions of space may be juxtaposed simultaneously as a gestalt and where minutes and hours are only a few seconds in duration (Joseph 1996, 2011a). Dream-time, and dream-consciousness, do not obey the laws of classical physics or Einstein's laws of relativity, but the quantum laws of "Multiple Worlds" and quantum physics.

Space-time is conceptualized as the 4th dimension (Einstein 1961). However, instead of a four-dimensional universe, three of space and one of space-time, there may be six dimensions, three of space, and three of time; i.e. space-time, conscious-time, and dream-time with all three time dimensions subject to contraction and dilation independent of each other. Compared with the 10, 11 or more dimensions of "string" and super-string theory (Kaku 1999; Polchinski 1998), six dimensions may be a conservative estimate.

These are just some of the conundrums of time travel.

2: Time Is Relative: Future, Past, Present Overlap and Exist Simultaneously

The Physics of Time

There is no universal now (Einstein 1955). Time is relative, and so too are the futures, presents, and pasts, which overlap and exist simultaneously in different distant regions of space-time. Time is relative, and the "present" for one observer, in one location, may be the past, or the future, for a second observer on another planet.

If an astronaut on Mars fires a gun at the "same time" as an astronaut on the Moon, a team of astronauts orbiting Mars will believe the gun on Mars was fired first, whereas observers on Earth will conclude it was first fired on the Moon. Even if both teams were supplied with atomic clocks, the clocks on the Moon, vs Earth vs Mars, would all show different times.

In 1971 Joe Hafele and Richard Keating placed atomic clocks on airplanes traveling in the same direction of Earth's rotation thereby combining the velocity of Earth with the velocity of the planes (Hafele & Keating 1972a,b). All clocks slowed on average by 59 nanoseconds compared to atomic clocks on Earth. Time, like the weather, is effected by local conditions. Under accelerated conditions and increased gravity, time slows down.

Atomic clocks tick off time as measured by the vibrations of light waves emitted by atoms of the element cesium and with accuracies of billionths of a second (Essen & Parry, 1955). However, these clocks are also effected by their surroundings and run slower under conditions of increased gravity or acceleration (Ashby 2003; Hafele & Keating 1972a,b); the same conditions which would enable a time traveler to accelerate toward the future.

As to what takes place locally, as observed by witnesses or recording devices in close proximity, it may be easy to agree if certain events occurred simultaneously or seconds apart--that is, what happened first, second, third, and so on. However, as distance between events and observers increases, what happens "here and now" cannot be compared to what takes place "there and then" or even "there and now."

Time has energy. As defined by Einstein's (1905b) famous theorem $E=mc^2$, and the law of conservation of energy and mass, mass can become energy and energy can become mass. Space-time is both energy and mass which is why it can be warped and will contract in response to gravity and acceleration (Einstein, 1914, 1915a,b; Parker & Toms 2009; Ohanian & Ruffini 2013).

Time is associated with light (Einstein 1961). Light has a particle-wave duality and travels at a maximum velocity of 186282 miles per second. However, time is not light, and light is not time. Rather, light can carry images reflected by or emitted from innumerable locations in space-time and can convey or transport information from these locations which may be perceived by an observer. Time is perceived by an observer and for much of human history has been measured by celestial clocks such as the phases of the moon, and the tilt and rotation of Earth and Earth's orbit around the sun which marks the four seasons and the 24 hour day (Joseph, 2011b). Time is a circle and may be segmented into years, months, weeks, days, hours, minutes, seconds, nanoseconds as measured by various clocks from sundials to atomic clocks.

Time is curved and the same appears to be true of space-time as conceptualized by Einstein (1911, 1913, 1915a,b) and others. And just as tremendous increases in gravity can shrink the height of a man, increases in gravity can shrink the circle of time thereby forcing more time into a smaller space. If the distance between 12 noon and 12 midnight contract, then it would take less time for 12 hours to pass; and it is this phenomenon, predicted by Einstein (1961; Einstein et al. 1923) and Lorentz (1892) which explains time travel.

"Clocks" are relative. Jupiter does not have a 24 hour day but a 9.9 hour day. On Mercury a single day is equal to 58 days and 15 hours on Earth. A day on Venus lasts even longer: 243 days on Earth. Time is relative and even atomic clocks can slow down or speed up depending on gravity and acceleration (Ashby 2003; Chou et al. 2010; Hafele & Keating 1972a,b).

One must not give "clocks" a god-like status. It is a mistake to say that "there is no duration of time other than what can be measured by clocks." To understand the nature of time, then measures of time must include human clocks, especially as observation and measurement can effect and alter local events within the quantum continuum (Heisenberg 1958); and that continuum includes time. Even atomic clocks are relative. There is no universal "now" even when measured by the most accurate clocks in existence.

Then there is biological time, conscious time, dream time, and these times are seldom the same. and depending on mood, emotion, or boredom, may be out of synch with clock-measured time (Joseph 2010a). And the same can be said of the innumerable locations in space-time each of which have their own "now" and where past, present, and the future overlap relative to observers on distant planets, stars, galaxies, and even cities just miles apart (Einstein 1955, 1961).

Consider, for example, the moon, the sun, and the stars up above. From a vantage point on Earth, the moon we see is actually the moon from 1 second ago. The sun we observe is a sun from 8 seconds in the past. Upon gazing at the nearest star, Alpha Centauri, the star we see is from years ago since it takes 4.3 light years for its light image to reach Earth. If you stand in front of a full length mirror, just 3 feet away, and since light travels at 1 foot per nanosecond, you see

Relativity, Space Time...

yourself as you looked 6 nanoseconds ago. You are staring into the past. You always look younger in a mirror than you are and what you see is the "you" from moments before. Mirrors are gateways to the past even though the image you see is experienced in the "present."

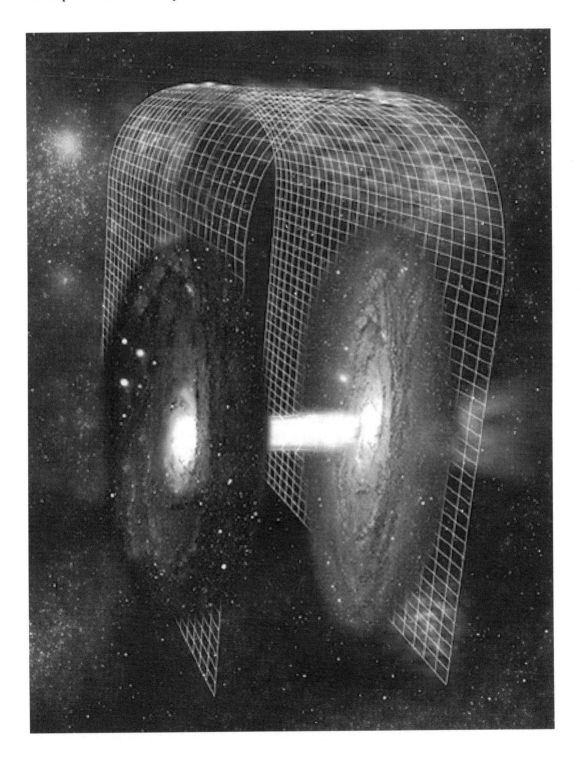

Quantum Physics of Time Travel

Space-time may consist of innumerable "mirrors" which tunnel into the past; referred to as Einstein-Rosen bridges, each of which may lead to a mirror universe on the other side (Einstein & Rosen 1935).

An image from the past, may be experienced in the present. Before the image from the past arrives it is still in the future, relative to the observer. A sunbeam which just left the surface of the sun will not arrive on Earth until the future, 8 seconds from now. Until it arrives, it is in the future, relative to those on Earth.

If an alien observer living on a planet in Alpha Centauri was gazing at Earth, then the present on Alpha Centauri overlaps with the past on Earth. The alien sees an Earth from 4.3 light years ago. The reverse is true for an observer on Earth. The past, future, and present, are relative to the observer.

The images of Alpha Centauri carried by light to Earth, are always of Alpha Centauri from 4.3 years ago. Hence, images of Alpha Centauri from 3.3 years ago, will not be received on Earth until a future date, one year from now. The future, or rather this particular future of Alpha Centauri from 3.3 years ago, exists in a distant location moving across space-time, and is speeding toward Earth and will arrive in one light year. The past can be the future and both may exists before they arrives in the present only to again become the past thereby creating a circle of time.

The entire prospect of time travel, as based on quantum physics and Einstein's theories of relativity, requires that "a" past and "a" future exist simultaneously, albeit in different distant locations and separated by varying distances all of which are relative and in motion in the dimension known as space-time. And just as the circle of time is that of a future which will becomes the present and then the past, the same is true of time travel: The future leads to the past, for it is only upon accelerating toward light speed and into the future that the time traveler can return to the past, but only if she exceeds the speed of light.

Past, present and future are relative, and also overlap. If past and future were not locations in space-time and flowing in various directions, one could not see images of stars from the long ago. If the time traveler could accelerate fast enough and overtake images of Earth mirrored in beams of light on their way to Alpha Centauri, he would see an Earth from the long ago; but only if he traveled faster than the speed of light.

If a space-time traveler, no matter what her velocity, were voyaging somewhere between Earth and Alpha Centauri she would encounter light-images from Earth's past before they arrive at some future date in Alpha Centauri. Once they pass her by, they are in her past but still in the future for those on Alpha Centauri. Likewise, she would come into contact with beams of light reflecting the images of Alpha Centauri before they arrived on Earth. Thus, from the perspective of an observer back on the home planet, the time traveler would encounter multiples futures--futures which have not yet arrived on Earth or Alpha Centauri. However, it is only upon accelerating beyond light speed that the Time-Traveler can catch up with light which has left Earth or Alpha Centauri; and this is because acceleration will shrink space-time, bringing the future closer to the present.

A similar principle applies to time travel. By accelerating toward light speed, space-time contracts (Lorentz 1982; Einstein 1961; Einstein et al. 1923), and the distance between the future and the present and distant locations in space time shrinks and are closer together.

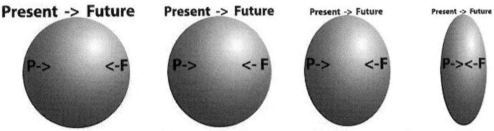

Contraction of Time and Space-Time

Figure: Contraction of space-time as velocities increase and approach light speed.

If a time traveler left Earth on a journey to Alpha Centauri, and quickly reached a velocity of 90% light speed, instead of taking 1569 days, the entire trip would be shortened to 685 days, from 4.3 years to 1.87 years. Because of the contraction of space-time, each day in the time-space machine at 90% light speed would propel him 2.29 days into the future of Earth. At 90% light speed, 2.29 days on Earth shrinks to 24 hours in the time machine.

When a time traveler accelerates toward light speed and journeys to the future, time in the world outside the time machine will be different from time inside the time machine, and the same is true of biological time and conscious time. Whereas time contracts and slows down in the time machine (relative to an observer on Earth), it speeds up on Earth (relative to the observer in the time machine). This is because the contraction of time-space, of distant locations, occurs not for all of time-space, but the time-space occupied by the time traveler.

H.G. Well's Time Traveler began the voyage through time in his laboratory and he could see events, people, and even a snail whiz by.

"The laboratory got hazy and went dark. Mrs. Watchett came in and walked... towards the garden door. I suppose it took her a minute or so to traverse the place, but to me she seemed to shoot across the room like a rocket."

Time, however, is relative. Although "Mrs. Watchet, from the perspective of the Time Traveler, seemed to race across the room, from Mrs. Watchet's perspective the Time Traveler would appear to be frozen in time or moving exceedingly slowly.

Quantum Physics of Time Travel

Well's Time Traveler also kept his eye on the laboratory clock and noted "a moment before it had stood at a minute or so past ten; now it was nearly half-past three!" A clock outside the time machine was therefore ticking away rapidly (relative to his clock inside the time machine), whereas his clock inside the time machine, from the perspective of Mrs. Watchett would run very slowly; exactly as predicted by Einstein's theories of relativity (Einstein 1914, 1915a,b, 1961).

When a time traveler accelerates toward light speed and journeys to the future, time in the world outside will be different from time inside the time machine, and the same is true of biological time and conscious time.

Consider for example, 30 feet of space which contracts to 10 feet. Those inside the time machine need only walk 10 feet whereas those outside the time machine must walk 30 feet. Likewise because the time traveler's clock runs more slowly, and since more time is contracted into a smaller space, it might take him 10 minutes to get 30 minutes into the future. By contrast, it takes those outside the time machine longer to get to the future because it is further away and as their clocks are running faster and it takes more time.

This is why a time traveling twin will age less than the twin left behind. Relative to each other, time and the aging process speeds up for the twin on Earth, and time and aging slows down for the twin in the time machine. Because time-space has contracted, and since it takes less time to get to distant locations which are now closer together, the time traveling twin arrives in the future in less time than her twin on Earth.

3: Time Dilation And The Contraction of Space Time

There are at least three dimensions of space and at least one dimension of space-time (Einstein 1961; Minkowski 1909). Time in one location is different from time in another location. The future, present, past, exist simultaneously in different, overlapping locations in space-time. Space-time can be twisted, folded, curved, and shrunk (Parker & Toms 2009; Ohanian & Ruffini 2013).

Space-time is elastic. Increases in speed increases gravity and mass which shrinks space-time causing space to curl up and contract (Einstein 1915a,b, 1961). For example, a 100 pound woman with a height of 5ft 2 inches on Earth, would weigh 235 pounds but shrink in height by approximately 16 inches if she was standing on a planet the size of Jupiter. Gravity pulls, it does not push. Likewise, increases in acceleration increases mass (as mass gains energy from the acceleration) which increases gravity. Increases in gravity cause space-time to shrink. In consequence, the distance between the "present" and the future decreases because of the shrinkage of the space between them. The present and the future are pulled closer together. Distant locations in space-time are no longer so far apart.

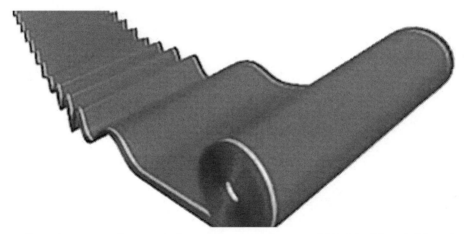

Imagine a newly married couple, a cowboy and his blushing bride who is sitting shyly on one end of a 10 foot carpet on a hardwood floor whereas he sits at the other end. The cowboy is thinking about what he is going to do to her in the very near future and he takes his rope and lassos her around the waist. Then he begins to pull her toward him and the carpet she is sitting on and which is between them begins to curl up and fold up, creating crests and valleys and a carpet warp. The far edge of the carpet, and his bride still sitting on top of it, are being pulled closer and closer to her husband. The cowboy husband is like the time traveler in the present, his bride is the future. But he does not move toward the future. The carpet warp caused the future to move closer toward the time traveler.

Space-time can be stretched, shrunk, and it may curl up in response to tremendous speed and gravitational influences, pressures and tensions (Carroll 2004); like wrinkled skin exposed to too much sun. Thus one layer of space-time may fold up and over another such that distant locations come to be side by side, or face to face; like the two handles on an accordion when the bellows are squeezed together; or two crests of a wave when moments before the sea was calm and the tips of the crests, when flattened out, were far apart.

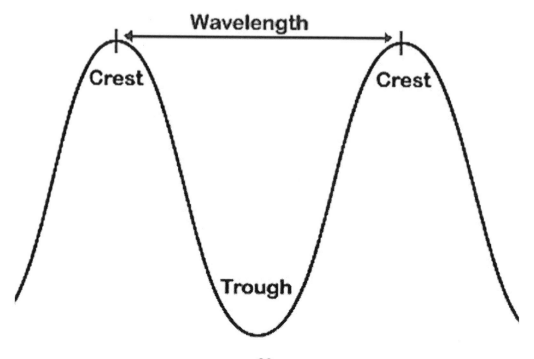

Relativity, Space Time...

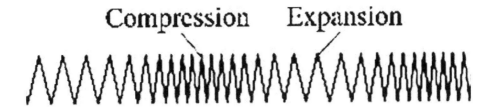

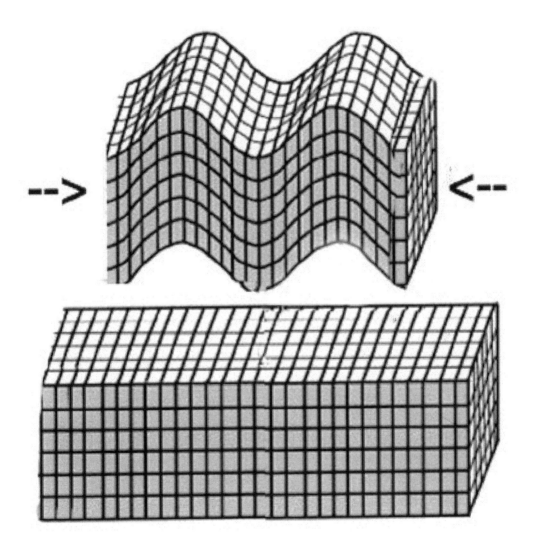

Because the future, present, and the past already exist simultaneously in different locations in space-time, it could also be argued that the time traveler accelerating toward light speed does not travel into the future, but rather because of this folding, contraction, and curvature of space-time, that "a" future location in space-time arrives more quickly (relative to his frame of reference). The contraction of space moves the future closer to him and it is no longer so far away. However, for the future to arrive more quickly means it must exist before it arrives.

The Future and the Past Are Relative

At 13.1 billion light years from Earth, a galaxy named "z8_GND_5296" is among the most distant galaxies so far detected. Although light, distance, and time are not synonymous, the fact remain that the light beams carrying the image of "z8_GND_5296" are reflections of this galaxy's past. However, each beam (particle-wave) of this light, before reaching Earth, is in the future from the perspective of observers on Earth, until the moment it is received and perceived on Earth; at which point it is in the present, and then continues into the past. Likewise, images of Earth carried by light on their way to "z8_GND_5296" are from our past but will not be received by "z8_GND_5296" until some future date on "z8_GND_5296" and thus represents a future event from the perspective of "z8_GND_5296" (as it has not yet been received). Events from the past on Earth take place in the future of "z8_GND_5296."

According to Einstein's theories (1905a,b,c, 1915a,b, 1961; Einstein et al. 1923), to arrive at a future location in space-time requires the size and length of space-time to contract; and this is what happens as velocity nears the speed of light: different locations in the future (e.g. futures which are 100,000 years apart, 1 million years apart, 13.1 billion years distant, and so on) become closer together so that it does not take as much time to get there; time being relative.

If the time traveler was journeying at 50% light speed to galaxy "z8_GND_5296" time-space would contract about 13% (See Table I). Hence, instead of taking 13 billion light years, it would only take 11.41 billon years to get there from Earth. At a velocity of 99% light speed, time-space would contract by about 80% and the trip would last 1.8 billion years. Like an accordion, space-time would be squeezed together and distant stars are no longer so far apart.

According to special relativity (Einstein 1905a,b,c, 1961), time can beat at different rates depending on velocity and acceleration. Time is linked to speed of movement. Space becomes compressed along the direction of motion (Lorenzt 1892, 1905; Einstein 1961; Einstein et al. 1923).

Space-Time Contraction

As one accelerates toward light speed space-time contracts and time slows down. Because the time traveler's velocity does not shrink, she can traverse the shrunken space more quickly than those left behind. Although space-time may curl up and shrink, her velocity, and the speed of light remains the same. Nevertheless, as the time traveler accelerates and increases velocity, she is catching up with the speed of light and thus the flow of time seems to be reducing speed relative to her own speed.

Relativity, Space Time...

Consider a train racing by at 100 mph when the observer is cruising along side in a sports car going just 30 mph. If the sports car accelerates to 150 mph, the train would seems to slow relative to the perspective of those in the sports car. The train will not be perceived as moving away as quickly. The distance between the train and the sports car will contract and decrease even though the train is moving at the same speed as before.

Likewise, speeding up and catching up with the speed of light causes time to slow down even though the speed of light does not change. Although the velocity of the train (and the speed of light), remains the same, the faster one goes, the slower time becomes; a function of accelerating velocity and the contraction of space-time. If the observer in the sports car accelerates to 200 mph, and because the speed of the train remains the same, the sports car will speed past the train and the train will fall further behind. The sport's car (and the time traveler) will arrive at a future location more quickly and in less time compared to those still traveling by train even though they are traveling the same distance.

Just as velocity and the space between the train and the sports car may contract or expand depending on their different speeds, time-space may expand or contract relative to different observers and their velocities; i.e. one in the time machine the other back on Earth.

Speed, that is velocity, per se is not effected by time travel. Velocity does not contract or dilate. Hence, since space-time contracts as one accelerates (and although time slows down), and as velocity is not effected then one can traverse and journey across this shrinking space more quickly, and cover the distance between the "now" and the "future" more rapidly because they are closer together.

Increases in velocity cause shrinkage not only of space time, but of the moving object (the time machine), which in turn increases the mass and gravity of that shrinking object which contributes to the collapsing and contraction of space time (Carroll 2004; Einstein 1913, 1914, 1915a,b).

According to Einstein's famous equation: $E = mc^2$, where E is energy, m is mass and c is the speed of light, mass and energy are the same physical entity and can be changed into each other (Einstein 1905a,b,c 1961). Because of this equivalence, the energy an object acquires due to its motion will increase its mass. In other words, the faster an object moves, the greater the amount of energy which increases its mass, since energy can become mass. This increase in mass only becomes noticeable when an object moves very rapidly. If it moves at 10% the speed of light, its mass will only be 0.5 percent more than normal. But if it moves at 90% the speed of light, its mass will double. And as mass increases it also shrinks and its gravity increases. This is because increased mass increases gravity which then pulls on the mass making it shrink toward the center of gravity.

Acceleration also increases G-forces, and centripetal acceleration increases gravity which is a major factor in time dilation and the contraction of space-

Quantum Physics of Time Travel

time (Carroll 2004; Einstein 1913, 1914, 1915a,b, 1961). Increases in gravity not only squeeze and crush those subject to these forces, but it can squeeze space-time and cause it to contract. Thus, increases in speed and gravity and g-force will slow time and shrink the time traveler and her time machine.

Gravity can also warp space-time and can cause space to curl up and contract, and this is because space-time has energy and thus an energy/mass duality as represented by the famous formula: $E = mc^2$. In consequence, the present and the future are pulled closer together. Distant locations in space-time are no longer so far apart; the result of increased speed and gravity.

Space-time dilates or contracts as gravity and acceleration increase. This is why at 90% light speed 2.29 days on Earth shrinks to just one day in the time machine and why 7 days in the time machine at this speed, would take the time traveler 16 days into the future. The distance between the present and the future has contracted so that the future arrives in 7 days instead of 16.

The Contraction of Time

Because the space-time machine, the time traveler, and any ticking clocks inside also shrink, the passage of time inside the machine also shrinks relative to time outside the machine. Inside the time machine time passes at the same rate because everything inside the time machine has shrunk to the same degree. Time is relative and it is only an outside observer at a safe distance from outside the time machine who will perceive that the time traveler's clock has slowed and the time traveler has grown smaller. By contrast, if the time traveler were to look outside the time machine it would seem that the outside observer has grown larger and his clock is ticking faster.

If a clock ticks faster, then more time will pass vs a clock which clicks slower which means less time will pass. Since distance between the "present" and the future decreases because of the shrinkage of space, and as his clock has slowed down then it takes the time traveler fewer clicks of the clock to arrive at the future vs those back on Earth.

The shrinkage of space-time has given rise to the famous "twin paradox" (Langevin 1911; von Laue 1913). If one twin leaves Earth and accelerates toward light speed, that twin will arrive in the future in less time than the twin left behind on Earth. Because it took less time, the time traveling twin does not age as much whereas the twin left on Earth ages at the normal rate. Hence, the time traveling twin will be younger.

The relationship between time dilation and the contraction of the length of space-time can be determined by a formula devised by Hendrik Lorentz in 1895. As specified by the Lorentz factor, γ (gamma) is given by the equation γ = , such that the dilation-contraction effect increases exponentially as the time traveler's velocity (v) approaches the speed of light c.

As detailed in Table 1, at 25% the speed of light, the effect is just 1.03,

Relativity, Space Time...

such that time and length slow and contract by 3%. At 50% light speed the factor is 1.15 whereas at 99% light speed, time is slowed by a factor of about 7. At 99.5% the factor is 10. At 99.5% of the speed of light, the time traveler's clock would be 10 times slower relative to a clock on Earth. If the time traveler achieves a velocity equal to 99.999% light speed, space shrinks by a factor of 224. However, since velocity remains constant the time traveler can cross this shorter length of space in a shorter period of time. Again, however, the shrinkage of time and space is relative to an observer on Earth.

Table 1 For each velocity, the time which elapses in the rest frame for each day measured by the ship's clock is depicted

v/c	Days	Years
0.0	1.00	0.003
0.1	1.01	0.003
0.2	1.02	0.003
0.3	1.05	0.003
0.4	1.09	0.003
0.5	1.15	0.003
0.6	1.25	0.003
0.7	1.40	0.004
0.8	1.67	0.005
0.9	2.29	0.006
0.95	3.20	0.009
0.97	4.11	0.011
0.99	7.09	0.019
0.995	10.01	0.027
0.999	22.37	0.061
0.9999	70.71	0.194
0.99999	223.61	0.613
0.999999	707.11	1.937
0.9999999	2236.07	6.126
0.99999999	7071.07	19.373
0.999999999	22360.68	61.262
0.9999999999	70710.68	193.728
0.99999999999	223606.79	612.621
0.999999999999	707114.60	1937.300
0.9999999999999	2235720.41	6125.261
0.99999999999999	7073895.38	19380.535
0.999999999999999	22369621.33	61286.634

Quantum Physics of Time Travel

Specifically, and if we accept that time and the speed of light are related, then if the time traveler journeys at 80% speed of light, then one day (1.197 days) from the perspective of the time traveler would be the equivalent of 2 days back on Earth. If he achieves 99% light speed, then 104 days in the time machine (time contraction) would be the equivalent of about 2 years on Earth. At 99.9% the speed of light, then 1 day (26 hours) in the time machine would be the equivalent of 6 years on Earth. If the time traveler wished to experience a future 2190 years distant she would have to spend one year in the time machine traveling at 99.999% light speed.

As detailed in Table 1, the amount of space-time contraction significantly increases as velocity nears light speed such that at lower speeds the effects are negligible even for velocities at 50% the speed of light. By contrast the amount of contraction becomes dramatic as velocities approach light speed. For example, at 99.999999% the speed of light, almost two years pass for every day in the time machine. At 99.99999999999 % of c, for every day on board, nearly twenty thousand years pass back on Earth. However, upon reaching light speed, time stops. It is only upon accelerating beyond light speed, that time runs backwards and the contraction of space-time continues in a negative direction. One must accelerate toward the future to reach the past.

Time-space contracts and so does the time machine which is flattened in the direction of motion. The time traveler and his vehicle shrinks like an accordion.

The paradox is that time speeds up outside the time machine, and slows down inside the time machine. However, from the perspective of the time traveler, time will not have slowed down but will appear to have speeded up back on Earth. Nor will space appear to have shrunk (from the time traveler's perspective). And this is because the time traveler and his time machine will also shrink and contract to the same degree and all is relative (Einstein 1961). By contrast, those on Earth will perceive the time machine as shrinking as it speeds up (assuming they can observe it). Eventually, the time traveler may shrink to the less than the width of a hair--at least from the perspective of outside observers.

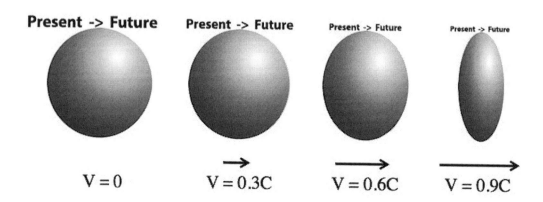

Relativity, Space Time...

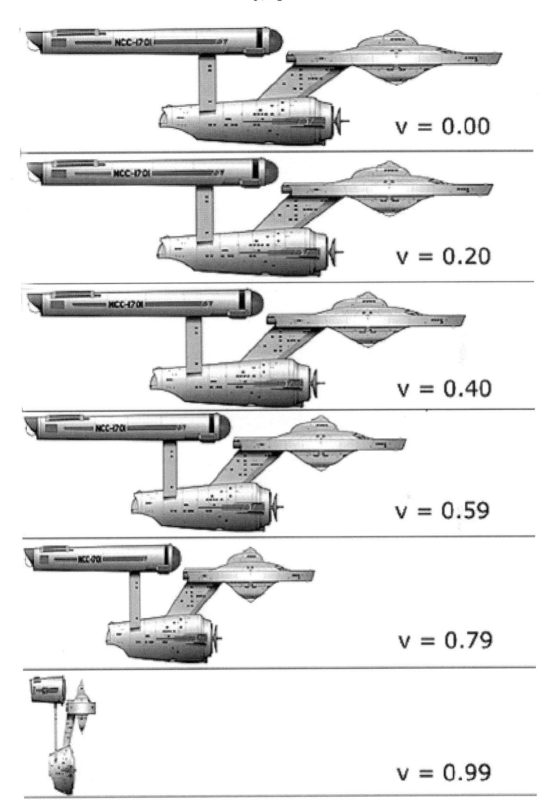

Likewise, the space between points in space-time may shrink to less than the width of a hair. Because velocity remains unchanged, the time traveler can travel between two points in space time more quickly because more time is also compacted and being squeezed into a smaller space, and the "now" and the "future" become closer together. Back on Earth, the distance between the future and the "now" remains the same and it take more time to arrive in the future. Therefore, the time traveler can reach the future more quickly than those on Earth.

What this all implies is that the future already exists, albeit in a distant location in space-time. Because of space-time contraction the time traveler does not really journey into the future. Instead the future journeys to him at an accelerated rate, and this is because the distance between the time traveler and the future shrinks as he accelerates and achieves a velocity of near light speed.

The future, or rather, multiple futures exist in various mobile locations in time-space, relative to the observer; and the same could be said of the past. There is no universal past, present, or future; all are relative and exist in multiple, overlapping locations in space-time.

4: Twins, Time Travel, Gravity And Aging

Because of time dilation and the contraction of space, once the time traveler lands on Earth, and depending on how fast and far into the future she is propelled, all her friends and relatives back on Earth may have died and a completely new generation of Earthlings may greet the time traveler upon her return. By contrast, since time slows down and time-space become squeezed together, the time traveler who arrives in the future may not have aged appreciably.

According to the well known "Twin Paradox," a thought experiment based on special relativity (Langevin 1911; von Laue 1913), because clocks inside the time machine run more slowly, a time traveling twin will age more slowly than her twin back on Earth. Thus, the Earth-bound twin will be be much older and may have already turned to dust if the time traveler arrived hundreds of years into the future. As summed up by Einstein (1911; see also Langevin 1911):

"If we placed a living organism in a box ... one could arrange that the organism, after any arbitrary lengthy flight, could be returned to its original spot in a scarcely altered condition, while corresponding organisms which had remained in their original positions had already long since given way to new generations. For the moving organism, the lengthy time of the journey was a mere instant, provided the motion took place with approximately the speed of light."

For example, say the time traveler is born in the year 2100, had a life expectancy of 80 years and would have died in the year 2180 if she had never left on her journey into the future. If she began her journey at age 20 in the year 2120, achieved 0.999999999999999 light speed and arrived (still breathing) in the future date of 2180, she would still have a life expectancy of 60 years (minus the 20 she already lived and time spent in the time machine). Upon arriving in the year 2180, she would still be 20 years old instead of 80 and could now expect to live another 60 years until the year 2240 (vs the year 2180 if she had never left home).

This premise is based on achieving near light speeds almost instantaneously and is supported by experiments with non-living, ultra-short-lived particles. For example, the muon particle is given a new lease on life when accelerated to a velocity of 99.92% light speed and its life span is nearly 25 times longer (Houellebecq 2001; Knecht, 2003). The muon particle not only lives longer but travels 25 times further thanks to its expanded life span. Particles, including phi mesons, which have been accelerated to velocities of 99.9% light speed also achieve significant life span extensions with a γ factor of around 5,000 (Houellebecq 2001). Presumably particles live longer because they have arrived in the future more quickly vs their counterparts traveling at their normal, slower speeds. Therefore, it could be predicted that a time traveler who journeys at near light speeds should also live longer compared to friends and relatives left back on Earth.

These same principles can be applied when traveling great distances across space to other stars and planets. If the journey takes place at near lights speeds, the space-time traveler may visit a distant star and then return home still fresh and young whereas her relatives and friends will have grown old and infirm and may have already died.

For example, if Gaia stays on Earth and her twin, Aurora travels at 80% the speed of light to Proxima Centauri which is 4.2 light years away, then Aurora's trip will take 5.25 Earth years (4.2/0.8 = 5.25). One day in the time machine at 80% light speed is equal to 1.67 days on Earth. Thus 1,916.25 days in the time machine (5.25 years) is equal to 3,200 days on Earth (8.76 years). Hence, Aurora's clock will tick 0.599% more slowly than Gaia's clock on Earth (5.25/8.76) and Aurora will age only 3.15 years during the journey (0.599 x 5.25 = 3.146) whereas Gaia will age 5.25 years. If Aurora immediately returns to Earth at 80% light speed she will be 6.3 years older and her twin will be 10.5 years older.

From Aurora's perspective, it was Earth and her twin sister Gaia, who traveled at 80% the speed of light; because time moves more quickly outside the time machine relative to inside. Thus Aurora is shocked (yet delighted) to discover that it was she, Aurora, who remained more youthful. Indeed, they are both shocked.

The differences in perspectives are because the twins did not have equivalent experiences. It is the twin who accelerates who experiences fewer ticks on her clock and more gravity and velocity. Likewise, the future and the past are not equal and are not equivalent and the same is true of gravity and light in the future and the past and in various locations in space-time. All are time asymmetric. Like the weather, time is not the same everywhere.

Because the time traveling twin vs the Earth-bound twin experience different gravitational conditions and as the time traveler experiences differential speeds of acceleration and then deceleration upon returning to Earth, the twins ceased to be "twins" and are no longer equivalent. The same would apply to "twin" atomic clocks (Hafele & Keating 1972a,b); the clock on Earth having more ticks than the clock on the time machine which runs slower.

Because of the increased velocity at which Aurora traveled, which in turn caused time-space to contact, thereby bringing the future closer to her present, she was able to reach the future in less time than Gaia since her velocity and speed of movement did not contract. Because space-time has shrunk, and since that space-time includes the future which is in a distant location, the distance to the future has contracted, and so too has time which has been squeezed into a smaller space. Thus, because the "now" and the "future" are closer together, and since velocity is not effected, it took Aurora less time to arrive at the future. Again, this implies that the future, or at least, "a" future must have existed before it became the "present" experienced by the time traveler.

From Aurora's perspective, time speeded up for Gaia, and this is because time and space-time were no longer equivalent for the two twins. Relative to her compressed, compacted point of view, everything outside the time machine is moving very rapidly, including space-time. Because Gaia's clock was speeding more rapidly, she experienced more time, more clicks of the clock, and aged faster than Aurora. Gaia experienced more clicks of the clock between the present and the future, whereas Aurora experienced fewer clicks of the clock and she aged less since for her, not as much time went by.

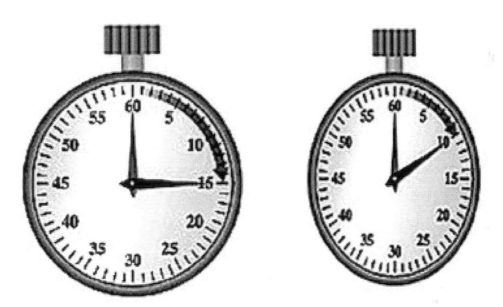

Figure: Because space-time contracts the future is closer to the present. Therefore it takes less time for the time traveler to arrive in the future. Although the twin arrived at the future at the same moment, it took Gaia longer since her clocked ticked more quickly.

Inertial Frames

From Gaia's perspective time slowed down for Aurora. And not just Aurora's clock, but Aurora would have appeared to be moving very slowly and maybe even frozen in time as she accelerated toward light speed. Aurora and her time machine would also appear to have shrunk in size; for its not just space-time which shrinks and contracts but the time machine and everything inside. By contrast, Gaia would appear to have grown tremendously in size and to be rushing by at lightning fast speed from Aurora's perspective. However, neither of these women would have any sense of how she appeared to her twin; and this is because their "inertial frames" were asymmetrical.

The size and motion of any object can only be described relative to something else. These are called frames of reference. If all frames are the same, then it would be impossible to determine size and velocity. There must be contrast, different inertial frames of reference. If everything was "wet" then we would have no conception of "dry."

In a non-inertial reference frame the laws of physics vary depending on the acceleration of that frame with respect to an inertial frame (Einstein 1961; Einstein et al., 1923).

If Aurora's velocity in the time machine was constant, her speed of movement in the time machine is zero. Although the time machine is speeding across time and the cosmos, since her speed is relative to the time machine then her speed is zero because she is not moving faster or slower than but at the same speed as the time machine. She does not feel any movement or motion or sense of going forward or backward except by contrasting herself with the surroundings outside the time machine. For example, if Aurora was riding in a train, she may look outside, down at the ground and can contrast the two frames of reference; her speed relative to the ground and her speed relative to the train. She is moving fast relative to the ground but is not moving at all relative to the train. However, what is "outside" may instead appear to be moving, whereas she may feel she is not since her speed relative to the train is zero.

If Aurora was riding in a Red Train and Gaia in a Blue Train and both trains were running parallel in the dead of night but at slightly different speeds, each woman would have a different inertial frame and might have difficulty determining who is going forward or backward. For example, if the Red Train is going 105 mph and Blue Train is going 100 mph and journey is incredibly smooth and the velocity constant, a passenger in the Red train whose windows only allow a view of the Blue train, might not realize the Red train is moving and may think the Blue train is going backwards (when it is just falling behind). Although the person in the Red Train is traveling at 105 mph relative to the ground, her speed is zero since her speed is relative to the train she is in. So, she is both moving (relative to the ground) and staying in the same place (relative to her train). Since she is staying in the same place (relative to the train) then her inertial frame is at rest. Because she is at rest, then it would appear to her that the Blue Train is in motion and is going backwards (when it is just going slower).

The inability to distinguish a resting state from uniform motion constitutes an "inertial frame of reference" and is based on Newton's first law of motion; an object moves at a constant velocity unless acted on by an external force. Therefore, an inertial frame is non-rotating, non-spinning, and there is no acceleration and velocity stays constant.

Relativity, Space Time...

Einstein's theory of special relativity, like Newtonian mechanics, assumes the equivalence of all inertial reference frames. According to Einstein's (1905a, 1961; Einstein et al. 1923) first principle of special theory of relativity, all physical laws have the same form in every inertial frame including the speed of light which, however, also stays the same in non-inertial frames. Whereas the speed of light may remain constant, size and length may contract and time may slow whereas from the frame of reference of the observer who is accelerating and shrinking in size as their clock slows, there is no sense of change. Aurora, therefore, would not sense that anything is different in the time machine. It would feel "normal." By contrast, Aurora's sister would appear to her as speeding up and growing in size. Gaia, however, would have exactly the opposite impression.

Aurora would also be unaware of any slowing of time within the time machine, and this is because she, as well as space-time, would have shrunk. The clocks inside the time machine and Aurora's biological clock, and her speed of consciousness would have been compressed and slowed to the same degree. From her point of reference, everything inside the time machine is normal whereas from the vantage point of those outside the time machine, Aurora is frozen in time, has shrunk in size and is moving very slowly.

Times-space contraction and the slowing of the time traveler's clock could be called the "Rip Van Winkle Effect" and could be likened to a state of semi-suspended animation, with each "hour" in the time machine corresponding to weeks, months, years, or centuries back on Earth depending on the near light speeds attained. When Rip Van Winkle wakes up, he finds himself in the future.

The Twins and The Past

And what would happen if a time traveling twin journeyed into the past? Would she age in reverse?

The only way to travel into the past, it to first travel into the future. Time passes by more swiftly as velocity approaches light speed, but at the moment the time traveler exceeds light speed, he will experienced a time reversal; space-time contraction simply continues in a negative direction and all clocks begin to tick in reverse. Likewise, the time traveler headed toward the past may age in reverse, becoming a teenager, then a child, infant, neonate, fetus, embryo, sperm and ovum. If this scenario is correct, then the time traveler could travel no further back in time than the day she was born, or before the time machine was built.

Or, the aging process may be accelerated and she may become an old as the time she visits. If she leaves the year 2200 and arrives in 2000, she may become 200 years old--at least compared to those back in the year 2200!

The backward journeying Time Traveler may disintegrate and decay as she journeys into the past. Decay processes are in line with what we know of the past--that which is from the past, decays. Memory of the past decays. Hence, the past decays. And the older that object is, the further back in time from the present, the more it has decayed.

Light speed must be exceeded to travel back in time. Clocks which had run increasingly slower as velocity reached light speed, stop at light speed, and upon exceeding the cosmic speed limit, these clocks run backwards and may increasingly run faster as superluminal velocities increase. Therefore, she may grow older more quickly, or younger more quickly, or she may self-annihilate upon encountering her double in the past; the problem of duality.

She who travels backward in time, already exists in the past. Unless the backwards in time traveling twin has negative mass and negative energy (a transformation which may occur upon exceedingly light speed), then this duality would violate the conservation laws of energy and mass. Upon encountering her self in the past, she and her double may self-annihilate. That is, just as space-time continues to contract in a negative direction upon exceeding light speed, the time traveler will contract in a negative direction and may consist of negative energy and mass.

Thus she may grow older more quickly, or age in reverse, or self-destruct if the negative energy/mass twin comes in contact with the positive energy/mass twin. The there is the problem of miniaturization--a consequence of length contraction.Traveling backwards in time poses many risks and may not be a successful means of finding the fountain of youth.

5: Time Travel And Aging: Clocks, Gravity, Altitude, Longitude, Longevity

So long as the time traveler is traveling at near light speed her clock and her mind are slowed and she ages more slowly compared to her twin on Earth. It is only when she reaches her destination in the future, and slows her velocity, that her clock and her mind speed up. From the perspective of an outside observer, the time traveling twin would be slowly emerging from her comatose state. It is only when her velocity decreases to a speed equivalent to the velocity of Earth that she awakens in the future only to discover that her sister has aged whereas she has remained young. Unfortunately, although she may be younger, she may not be very healthy.

The "Twin Paradox" as based on Einstein's theories, predicts that relative to the time traveler, the twin back on Earth will age more quickly. The clock in the time machine runs more slowly, whereas for those back on Earth, clocks run faster and they age accordingly. For example if the time traveler left on his travels at age 20 and journeyed for 7 days at 99.999999999% the speed of light, his twin back on Earth would have aged 42 years and would be 62 years of age vs 20 years and 7 days for the time traveler who would have only aged 7 days. But is it true that time traveler would have only aged by a few days?

If the clock on board the time machine slows down, does one's biological clock also slow? Does the heart beat slower? Is it possible that the stress of time travel might cause premature aging?

It has been demonstrated that atomic clocks at differing altitudes will eventually show different times; a function of gravitational effects on time. The lower the altitude the slower the clock, whereas clocks speed up as altitude increases; albeit the differences consisting of increases of a few nanoseconds (Chou et al. 2010; Hafele & Keating, 1972; Vessot et al. 1980). "For example, if two identical clocks are separated vertically by 1 km above the surface of Earth, the higher clock gains the equivalent of 3 extra seconds for each million years (Chou et al., 2010). The speeding up of atomic clocks at increasingly higher altitudes has been attributed to a reduction in gravitational potential which contributes to differential gravitational time dilation.

A predicted by Einstein, clocks run more slowly (time contraction) near massive objects whereas time dilates and runs more quickly as gravity is reduced. Increases in altitude and reductions in gravity speed up the clock, whereas decreases in altitude and increases in gravity slow the clock down.

Quantum Physics of Time Travel

There is more gravity close to the surface of the Earth and gravity decreases with increasing elevations above the center of gravity. For example, at 30,000 feet (9,000 meters) above sea level, one's weight would decrease by 0.29% (National Physical Laboratory, 2007) whereas clocks run faster. By contrast one is heavier at sea level and clocks run slower than at higher elevations since the gravity of Earth causes clocks to lose one microsecond every three hundreds years. Since elevations in altitude increase the distance from the Earth's center, gravity decreases. Decreases in gravity effects the passage of time thereby causing the clock to speed up (Hafele & Keating, 1972; Vessot et al. 1980).

Acceleration, however, expands mass (as energy is converted to mass) and increases gravity. Increases in gravity not only squeeze and crush those subject to these increases, but it can squeeze space-time and cause it to contract. Thus, increases in velocity and thus gravity and g-force will slow time and contract space-time, such that there is more time in a smaller space. However, contrary to Einstein's predictions, the time traveler may age more rapidly precisely because of the forces which make time travel possible.

Gravity Kills

A time traveler when accelerating toward the speed of light, and then shrinking and expanding in mass, would experience increases in gravity (due to increases in mass and acceleration). Gravity increases no matter how far from Earth if a vehicle is accelerating. And yet, increases in gravity, although slowing time do not slow the aging process, but speed it up. Those who dwell at higher altitudes and under reduced gravitational influences (which speeds up the clock), are healthier and live longer than flat-landers who experience greater gravity (Ezzati et al., 2011; Baibas et al., 2005; Winkelmayer, et al., 2009).

In one study it was found that men living at a mean elevation of 5,967 feet above sea level lived between 75.8 and 78.2 years which is 1.2 to 3.6 years longer than men dwelling at sea level who are thus closer to the center of gravity. And the same was found for high altitude dwelling women who lived on average from 80.5 to 82.5 years which is 0.5 to 2.5 years more than females residing at sea level (Ezzati et al 2011). In yet another study it was found that regardless of education, weight, smoking, alcohol consumption, blood pressure, total cholesterol, blood sugar, blood fats, and uric acid, death rates from all causes were 43% lower among men, and 31% lower among women, in the mountains than in the lowlands (Baibas et al.k 2005).

Of course, it could be argued that high altitude longevity and health has nothing to do with time dilation or gravity, but is a function of numerous other variables such as the lower oxygen environment and increased exposure to solar radiation which speeds up the synthesis of vitamin D. As for gravity, a person flying 40,000 ft above sea level, will feel more gravity when they pass over a mountain than when they fly over the open sea (National Physical Laboratory,

2007). On the other hand, those standing on the mountain feel less gravity because they are further away from Earth's center of gravity. For example, a person weighing 150 pounds at sea level, would weigh approximately 149.92 pounds at 10,000 feet above sea level.

Thus, life span expands and atomic clocks increasingly run faster as elevation and gravity decreases. By contrast, the time traveler would experience increases in g-forces (at least as they accelerate toward the speed of light) and a slower clock as well as radiation. As based on these findings it could be predicted that the time traveler might age more rapidly, suffer more health problems, and die sooner as compared to the twin she left behind. Yes, perhaps she would be younger, but she would have a lot of health problems and might die quicker.

Longevity and Latitude

It appears that time is not on the side of the time-traveler. Just as lowlanders die more quickly than high landers, those at higher/lower polar latitudes (and who are subject to increased gravity and centrifugal forces) suffer more serious cancer-related health problems than those dwelling in the equatorial regions (Grant, 2010).

Gravity and centrifugal forces differ according to latitude. As one approaches the polar latitudes gravity increases, a function of outward centrifugal forces which are larger at the polar regions vs the equatorial regions due in part to the changing circumference of Earth and the equatorial bulge. Sea-level gravitational acceleration increases from about 9.780 m·s^{-2} at the Equator to about 9.832 m·s^{-2} at the poles, so a person will weigh about 0.5% more at the poles than at the Equator (Boynton 2001; National Physical Laboratory, 2007). Thus, gravity and centrifugal forces are weaker at the equator.

Likewise, whereas increases in elevation and altitude increases life expectancy and health, decreases in latitude reduces the incidence of severe health problems such as cancer, at least among non-human animals; i.e. mammals die younger and get sicker as distance from the equatorial regions increases and proximity to the polar regions decreases (Grant, 2010; Mead 2008). Thus, increases in gravity secondary to increases in centrifugal forces may not good for one's health, at least in respect to cancer; and it is conditions such as these which the time traveler must deal with.

The time traveler who journeys toward the future will experience g-forces and time contractions completely unlike those dwelling on Earth no matter what the altitude or latitude. Unfortunately for the time traveler, it is precisely because of these latter variables that the time machine may instead become a killing machine and propel him to an early death.

6: Acceleration, Light Speed, Time Travel, G-Forces And Warp Drives

To journey to the future or the past requires a journey through space.

Earth, and our solar system are in motion, with Earth having a solar orbital speed of 108,000 km/h (~70,000 mph) and our solar system a speed of 720,000 km/h (450,000 mph) as it circles the Milky Way galaxy. If a time traveler stepped into a time machine in Los Angeles, London, or Beijing, and then set out for the future, he would find himself alone in space millions of miles in front of this planet as it orbits through the cosmos.

What if one were to create a time machine which remains anchored and tethered to Earth? Presumably, under these conditions, everyone and everything on Earth would also be propelled to the future at the same rate as the time traveler. All clocks would run basically at the same speed. The passage of time, time being relative, would not change from what is experienced as the normal or average flow of time. The time traveler might as well be sitting on his couch. Unless the time machine is constructed of Plank Length black holes with negative energy and mass, or unless some means is found to shrink the time machine to a size smaller than a Planck length (a regions of space where time-reversal is most likely to occur) the time traveler cannot be anchored to Earth.

Time machines must be capable of voyaging through space. Unfortunately, the space-time traveler will experienced prolonged periods of injurious g-forces while accelerating toward light speed. Unless willing to tolerate high levels of g-force for days, weeks, months or years, time travel into the distant future is just not practical. For example, it would take a time traveler, accelerating at a rate of 1 g (the gravity of Earth) 250 light years to reach a speed of 99.9992% of c.

In order to travel to the past or the future of this planet, and because Earth and this solar system are in motion, one would require a time machine that could also serve as a technologically advanced space craft with the power and capability of accelerating toward the speed of light, counteracting g-forces, maintaining velocity, and then decelerating and locating and landing upon an Earth which exists in a distant future or past location in space-time.

Time in Time-Space Travel

If the time traveler journeyed at a velocity faster than the movement of Earth, then, according to Einstein's theories of relativity (1961; Einstein et al.

1923), space-time would contract around him and his clock would run slower than the clocks of those back on Earth and he would age more slowly as compared to his counterparts back on Earth. Of course, if absent protective life-support or a space capsule, the time traveler would quickly be reduced to a lifeless corpse.

It has been demonstrated that atomic clocks on planes run slightly faster with respect to clocks on the ground (Hafele & Keating 1972a,b). By contrast, the gravity of Earth causes clocks to lose one microsecond every three hundreds years (Chou et al. 2010). Likewise, global positioning satellites (which provide location and time information to those on Earth) also experience gravitational time dilation and their clocks must be regularly corrected (Ashby 2003). The same is true for the International space station (ISS).

The International Space Station, where the gravitational strength is 89% of the Earth's surface has an effective gravity (gravity plus acceleration) close to zero, in part because it is in "free fall" despite its horizontal speed. The ISS also falls about 20 meters with every orbit. However, because of centrifugal forces which pull the ISS away from the center of Earth's rotation thereby countering Earth's gravity, the ISS remains in orbit at the same distance above the planet. And yet, as it falls (but does not fall) and as its speed of acceleration and velocity exceeds that of Earth, space-time slightly contracts and astronauts aboard the ISS are continually propelled toward the future.

NASA astronaut Ed Lu using a traditional clock has estimated that for every 6 months aboard the ISS, astronauts will age 0.007 seconds less than those on Earth (Lu, 2000).

The ISS orbits above Earth approximately 15 times per day at an altitude ranging from 330 km (205 mi) to 435 km (270 mi) which is adjusted via reboost manoeuvers using the engines of the Zvezda module or visiting spacecraft. The ISS travels at an average speed of 27,724 kilometers (17,227 mi) per hour (4.78 miles / 7692.7 m per second) (NASA 2008). With an average velocity around 7700 meters per second then it is traveling about 0.0000257 the speed of light. This yields a time contraction factor of 1.00000000033. Hence, for each second on board the ISS, 1.00000000033 seconds pass on Earth, and for each year on the ISS time will have slowed by .0017344 compared to Earth time. Thus, it could be said that astronauts aboard the ISS experience time contraction and that the future arrives more quickly for them vs those on Earth.

Gravity and Health in Space Travel

Is it really true that astronauts and time travelers would age more slowly as predicted by Einstein's theories and as supported by astronaut Ed Lu's anecdotal evidence? An astronaut hoping to visit the ISS must first escape Earth's gravity and experience increased g-forces. A space-time traveler will experience increased g-forces for days, months, or years. Time travel would subject the body to tremendous stress.

For example, although reductions in gravity, at least when comparing mountain dwellers from those who reside at sea level, is associated with better health and a longer life, it is well known that microgravity is not good for one's health, resulting in decreases in muscle mass and bone density (Levin & Schild 2010). Likewise, increases in gravity, as measured in g-force, is a killer.

Four separate studies examining health and longevity including those comparing astronauts with men and women of similar age, race, and educational background, as well as pilots and those working in dangerous industries, all found the same: Astronauts die at a younger age including from cancer (Hamm et al., 1998, 2000, Logan 2003; Peterson et al., 1993). Astronauts spending 6 months or more on the ISS do not emerge back on Earth looking younger and healthier, but more like old men and women who have difficulty standing, walking, maintaining their balance and lifting heavy objects (Levin & Schild 2010).

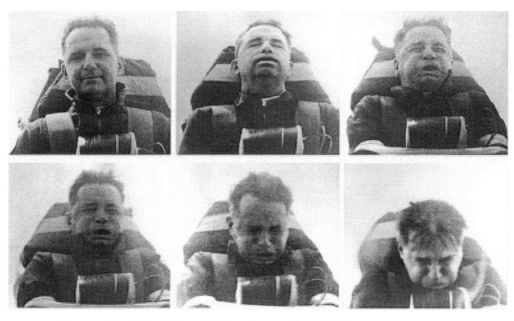

Figure: Test Pilot Experincing Increasing G-Forces

G-Forces And Health in Time-Space Travel

Astronauts also experience g-forces unlike most folks back on Earth; a consequence of acceleration to escape Earth's gravity and when changing velocity. A g-force of 1 is equal to Earth's gravity. However, during rocket launch astronauts may experience up to 3 gs, but because of the construction of the seating which spreads the g-forces across the body and offers protective support, they can tolerate these stresses without difficulty for several minutes (Levin & Schild 2010). Nevertheless, after a few weeks in space, astronauts are so weakened they must wear protective G-suits during re-entry to prevent blackouts (Levin & Schild

2010. The duration of these 2-3 g-forces experienced during acceleration and lift off, deceleration and reentry, and maneuvers in space, in total, may range from 15 to 20 minutes(Levin & Schild 2010.

The time traveler upon accelerating toward the speed of light would experience hypergravity and increased g-forces for years even if the target speed is just 80% light speed. G-forces are not good for one's health or longevity (Elert, 2004; Froklis et al., 1997; Oyama 1982, Pitts et al. 1975).

The term "g-force" is a measure of acceleration. At normal acceleration, i.e. the movement of Earth, the standard is 1g. Everyone feels "normal" at 1 g, weightless at 0 g, but twice as heavy at 2 g. Experienced fighter pilots can withstand acceleration of up to 9 gs for up to 1 second, after which they black out or die because of the acceleration induced pressure crushing the body and its internal organs. A human weighing 200 pounds and subject to 9 g's for just a one second will feel as if 18 thousands pounds of crushing pressure were pressing against their entire body, making it impossible to breath or for the heart to pump blood or for blood to circulate.

Death is the outcome if subjected to 6 g's for more than a few seconds, or 5 g's for more than a few minutes, as has been demonstrated with a variety of mammals including rats and dogs (Froklis et al., 1997; Oyama 1982, Pitts et al. 1975).

Even prolonged exposure to 4.1 gs increases mortality and causes significant injury to the body; and this is because increased g-forces causes stress, injures internal organs, and increase the metabolic and energy requirements such that body fat is quickly used up. After body fat is digested and turned into energy, muscle mass is consumed for energy; the body begins to digest itself (Oyama 1982, Pitts et al. 1975).

Given current and foreseeable future technology, it is not possible for a human, living on Earth, or orbiting this planet in a time machine or space capsule, to instantly achieve light speed. To accelerate toward light speed means increased g-forces. It is impossible to escape the ravages of prolonged G-forces if one wishes to journey to the future. Even if it were possible to achieve the speed of light in just a few seconds, a few minutes, or even a few hours, any living organism would be reduced to a flattened smear of micro-molecular ooze the thickness of a Planck Length by the incredible g-forces generated during acceleration. For example, if the time traveler accelerated to about 55% of light speed, i.e. 100,000,000 m/s in just 10 seconds he would experience millions of pounds of pressure ((100,000,000 m/s)/10's = 10 million m/s^2 = 1 million g), thus killing him instantly.

Even if accelerating at just ~250 m/s^2 with the goal of initially achieving a speed of 62,000 miles per second (0.33 the speed of light, c/3 or 100,000 km/s) a 200 pound time traveler would be exposed to 25 gs for (100,000,000 m/s)/250 m/s^2 =) 400,000 seconds (= 6,666 minutes = 1,111 hours = 4.6 days) and they would be crushed by over 5,000 pounds of pressure. Thus, for 399,999 of those seconds they would already be dead.

Research has demonstrated acceptable survival rates at 3 g in mammals (Froklis et al., 1997; Oyama (1982; Pitts et al. (1975), which is also the average g's experienced by astronauts on and off for about 15 minutes of each mission. However, if accelerating toward the speed of light at 3 g, it would take the time traveler approximately 31 years to reach 99.99999999999 % light speed. Once at 99.99... the time capsule would no longer need to accelerate, g force would disappear, and for every subsequent day in the time machine approximately 20,000 years would pass by on Earth.

Unfortunately a 20 year old time traveler who took 31 year to achieve near light speed would be 51 years old once near light speed was achieved. Given the incredible g-forces and all attendant stresses experienced, it is unlikely they would be in good health, and more likely they would have aged beyond their years and would arrive in the future in need of a coffin.

Even if the time traveler survived those 31 years at 3 g, they could not simply stop and land on Earth 20,000 years from now. They would have to decelerate and slow down at 3 g. This would take another 31 years to decelerate from 185,999.999 miles per second to around 185 miles per hour which is the average landing speed for a fully loaded 747 jetliner.

The time traveler, however, would also have to turn around, if he or she wanted to land on Earth. And, the distance they would travel in order to slow down from near light speed, with a declaration rate generating 3 gs, would be in excess of 25 billion miles. That is, if they were coming in for a landing, they would have to slow down to 185 mph from near light speed, and this slowing descent from an initial speed of 185,999.999 mph with a maximum impact of 3 gs would require a "runway" 25 billion miles long. Therefore, the time-traveler, if their destination was Earth, would have to first overshoot their target by 25 billion miles and then turn 180 degrees at light speed (which would generate incredible g forces) and then aim to intercept a still future Earth which will be at least 25 billion miles distant and thousands of additional years in the future. Unfortunately, once they step out of the time machine 20,000 years from now, the time traveler will be over 82 years old and most likely will need a wheel chair or a stretcher, if not a coffin.

Fuel and G-Forces

It would take a time traveler, accelerating at a rate of 1 g (the gravity of Earth) 250 light years to reach a speed of 99.9992; but once at this velocity, and due to time contraction she would age only an additional 6 years by the end of her journey when she arrives in 250 light years into the future. However, the time traveler would still have to slow down to zero velocity so as to make landing. This would take another 250 light years, and she would age another 6 years and would have thus aged at least 12 years but would arrive 500 years in the future.

Although accelerating at 1 g would make for a comfortable journey, there remain the problems of designing an engine powerful enough to accelerate toward light speed. Then there is the problem of a power source and providing the fuel for the engine. The Saturn V rocket launch vehicle, service module, and lunar module which propelled men to the moon contained 5,625,000 pounds of propellant (960,000 gallons), enough to fill 54 railroad tank cars; and each F-1 engine burned 3,357 gallons (12,710 liters) of propellant every second.

Fully fueled, the Saturn V weighed 6.5 million pounds (3,250 tons) and reached a velocity of 7 miles per second (42,500 mph). The time machine would have to reach a velocity of 185,900 mi/sec to obtain near light speed. Even if the time machine weighed less than say, 1,000 tons, the fuel requirements alone would make the journey near impossible.

Warp Drives

As technologies develop and science marches on, these latter problems may be overcome. For example, it may become possible to harness anti-matter to combust matter, thereby releasing tremendous amounts of gamma ray photonic energy as they self-annihilate.

Yet another means would be to employ circular vacuums around the exterior of the craft containing pockets of positive energy and negative energy which would then repel (negative) and attract (positive) simultaneously. The positive energy would chase (be attracted to) the negative energy which would be repelled. Therefore, the positive and negative would be chasing and circling each other at increasing velocities, thereby giving the time-machine tremendous thrust (see chapters 12, 14, 15). A time machine constructed in this fashion might be saucer or oval shaped, with the negative-positive vacuum tubes circling the outer and portions of the craft.

Or it may become possible to exploit the incredible amounts of energy contained in spaces less than a Planck length ($1.61619926 \times 10^{-33}$ cm), the smallest measure of size, in which case, a trip 1000 years into the future might require less than 1000 pounds of highly condensed energy. In fact, like space-time, as the time machine accelerates it would increasingly contract, such that at near light speed it may become smaller than a Planck length and blow a hole through time-space (see chapters 9, 10, 11, 12) thereby allowing it to tunnel through to the past, present, future, and any mirror universe which may lay on the other side.

7: The Curvature of Space-Time: Gravity and the Bending of Light and Time

According to Newton's theories, gravity acts as a force which can effect distant objects. Therefore, Earth and the other planets orbit the Sun because of the force of the Sun's gravity which maintains an instantaneous and continual grip on its satellites. Einstein rejected this view. According to Einstein (1905a, 1907, 1911, 1915a,b, 1961), nothing can have an instantaneous effect on a distant object since nothing can travel faster than the speed of light. Instead, Einstein asked: if gravity is not a force, how can it effect the orbits of planets and the objects and people living on that those planets, such as Earth?

Einstein reasoned that since the information or force acting on distant objects must be conveyed through space then it must be perturbations and alterations in the geometry of space which effects distant objects. It is the contiguity of distant objects connected by space-time which accounts for instantaneous effects. That is, it is the geometry and shape of space which is effected by gravity, which effects distant objects, including Earth. Distant objects are connected and linked by the geometry of space (Carroll 2004; Einstein 1915a,b 1961).

As theorized by Einstein (1906b, 1911, 1913, 1914, 1915a,b), the presence of matter causes a depression in space-time, warping the geometry of space-time. The presence of matter causes space-time to become non-Euclidian by creating deformations in space-time geometry. It is these structural changes in the geometry of space-time which informs objects and planets, at a distance, of any changes in gravity or the mass of objects and their gravity. Matter acts on space-time and alterations in space time act on matter.

The logical assumption is that space-time must consist of substance. That substance is energy, and energy can become mass as dictated by Einstein's famous formula $E=mc^2$.

Movement through space, and acceleration produces energy which is absorbed by the time machine and is transformed into mass. As the time traveler accelerates, greater amounts of energy will be absorbed thereby increasing the mass and thus the gravity of the time traveler and his machine.

The mass of the time machine and time traveler is also compressed as they accelerate. The increasing mass and gravity of the time machine can also effect and warp the geometry of time-space.

Energy can become mass. Mass can become energy. Mass is a source of gravity. A swinging pendulum on a clock is heavier than a pendulum at rest, because kinetic energy gives the swinging pendulum additional mass. An electron whirling at near light speeds inside an LEP accelerator weighs 200,000 more than an electron at rest (Houellebecq 2001). However, as objects in motion gain mass they become heavier and this retards velocity which makes it harder to maintain speed or to accelerate. On the other hand, acceleration and increased mass also slows and retards the flow of time and alters the geometry of space-time, causing it to shrink (Carroll 2004). The future and the present become closer together. Even with a reduced speed the time traveler can still rush to the future in less time than those back on Earth.

Since the time traveler and his time machine are comprised of matter their presence and movement through time-space will also warp and depress the geometry of space-time thereby creating local and distant effects; and this includes shrinking and slowing time.

A clock will tick more slowly as it approaches the source of gravity. A clock on a planet with the gravity of Jupiter (24.79 m/s^2) will run slower than a clock on Earth (gravity = 9.78 m/s^2). Clocks on Earth (and Jupiter) run slower than a clock on Mars (gravity = 3.711 m/s^2) or the moon (1.622 m/s^2). Gravitational time dilation has been confirmed in a number of experiments (Chou et al. 2010; Pound & Rebka, 1959). As predicted by Einstein's equations, the gravity of Earth causes clocks to lose one microsecond every three hundreds years.

Gravity can shrink time and cause curvatures in space (Carroll 2004; Einstein 1961; Parker & Toms 2009; Ohanian & Ruffini 2013). Therefore, the increasing gravity and mass of the time machine will also contract and curve surrounding space-time. However, it will not effect all of space-time, only local space-time, the extent of its effects depending on its gravity/mass; and this is also true of different planets, stars, and galaxies.

The curvature of space-time is described by a mathematical quantity, the "Riemann curvature tensor" (after Berhard Riemann). In 1854, Riemann gave what became a famous lecture about gravity and the curvature of space. Einstein reformulated Riemann' theories which are now known as general relativity (1915a,b, 1961). The basis of Riemann and Einstein's work is the metric tensor, a collection of 10 numbers defined at each point in space. As dictated by these 10 independent tensor equations, space-time is curved differently at specific locations, a function of mass-energy density, pressure, stress, momentum density and energy flux in each location.

According to Riemann (1854), the warping, bending, and curving of space gives rise to what appears to be force--but according to Einstein (1911, 1913, 1914, 1915a,b, 1961) these forces do not exist. Einstein did not consider gravity to be a force, but as part of the geometry of space. The curvature of space is determined by the amount of energy-matter contained in that space (Mach's principle). Gravity is related to mass and acceleration. Mass can be converted into energy, energy into mass, and the total amount of mass (times c^2) plus the amount of energy is a constant quantity. Thus space is curved and bent throughout the universe, and these curvatures and twists and bends differ in various regions of space-time depending on the gravity and mass of different planets, stars, and galaxies. The universe, according to Einstein's theories, would not be a perfect sphere, but lumpy with innumerable cavities and folds created by the mass, energy and gravity of various stellar objects.

The Circle of Time

The universe is curved, at least, according to Einstein. However, if correct, this implies that just as traveling in a straight line on Earth will bring the traveler full circle to his starting point, the same could be applied to universe, light, and time. That is, not just space would be curved, but light, as well as time.

Time may be a circle, a cosmic clock which ticks at different speeds depending on the geometry of space-time. A logical corollary of this is what has been referred to as a Gödel universe (Buser, et al., 2013; Gödel 1949a,b, 1995; Rindler 2009) where there is no "absolute time" and where time circles back on itself in endless loops thereby making it possible for a time traveler to journey in a circle back to the point in space and time from where he started, even returning to the point where he was born.

As determined by the Riemann Einstein equations, the bending and curvature of space is directly correlated with the amount of mass and energy in that particular region of space. Variations in the distribution of matter and energy can generate almost every conceivable geometry in space-time and that geometry is relative to the amount of mass and gravity in different locations (Carroll 2004).

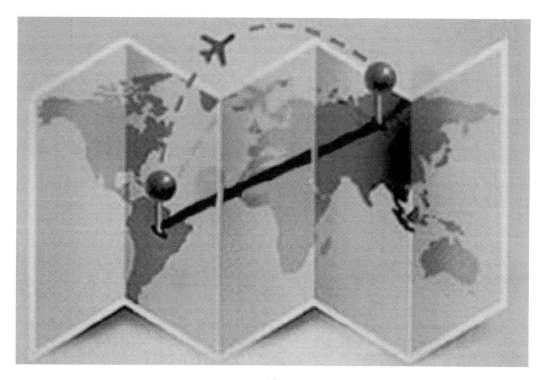

Relativity, Space Time...

In relativity, with sufficient concentration of energy and matter, space-time bends and contracts to such a disagree that distant locations can curl up right next to each other making time travel inevitable. Because of this bending, curling, and contraction, the distance between the future and the present shrinks; like two distant points on a map which are folded up next to each other.

For example, the distance between between Beijing China and Buenos Aires Argentina, is 12,326 miles. However, if these two cities were on a space-time map and that map was folded up such that the two cities were on two adjacent folds right next to each, then the distance, in a straight line going from fold to fold, would be much shorter.

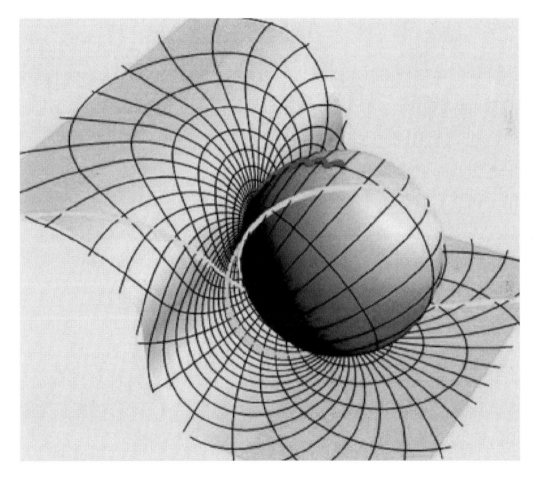

Gravity Holes in Space-Time

Einstein's curved universe is twisted and torqued as galaxy distribution is asymmetric, with great "walls" of galaxies throughout the cosmos which have clustered together. This clustering of billions of galaxies and vast reaches of empty space between them contribute to the unequal distribution of gravity, which causes space-time not just to curve, but to fold and curl up and to asymmetrically effect the flow of time.

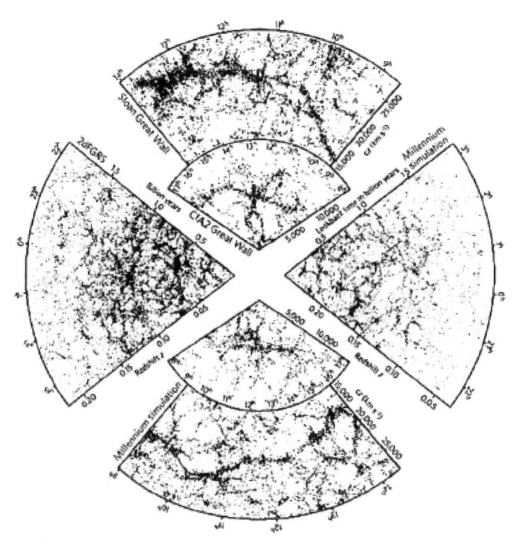

Figure. The diagram above is a depiction of the "Hubble Length" visible portion of the universe from the vantage point of Earth. Each black dot represents a galaxy. There are dozens of massive clusterings of galaxies creating "great walls" which are billions of light years in length.

Pockets, Warps, And Holes in Space Time

If sufficiently strong, gravity can even force space-time to curl up and fold over itself forming layers and even gravity holes in space-time (Bruno, et al., 2001; Eisberg & Resnick 1985; Garay 1995; Joseph 2010b; Smolin, 2002) which tunnel through the folds and the vacuum between them. Gravity, if sufficiently powerful, can cause distant locations in space-time to be pulled so close together they end up side by side or one on top of the others, like the crests of two waves or a carpet which has folded up and over itself. These curvatures, in turn, alter the trajectory of light (Einstein 1905b,c, 1911, 1915a,b). Light follows the curvature of space-time. When light passes near a massive object, it is bent and will curve, following the curving geometry of space (Parker & Toms 2009; Ohanian & Ruffini 2013; Thorne & Hawking 1995). Presumably, the same is true of time.

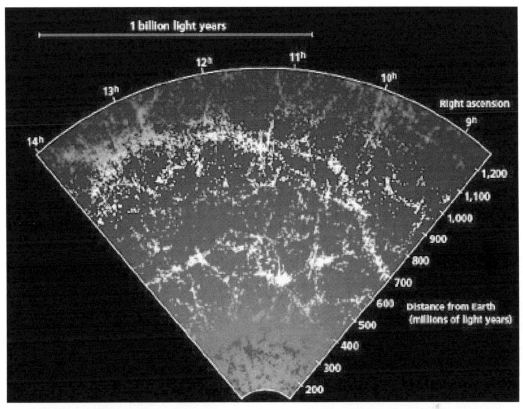

Figure. (above) Great galatic clusters forming galactic "walls" of galactic gravity over 1 billion light years in length

Quantum Physics of Time Travel

Imagine two hard spheres (or bowling), representing two galaxies, one (Sphere/Galaxy A) with a weight of 16 pounds, a second (Sphere/Galaxy B) with a weight 6 pounds. Each of these spheres are laid out on a thin 16 by 16 foot sheet of rubber (representing space-time) suspended in air at a distance of 6 feet between them.

The weight/mass (gravity) of Sphere/Galaxy A will causes a deep curvature (depression, or pocket) within the rubber sheet (representing space-time) and will drag surrounding space-time toward it and into the depression. Sphere/Galaxy B has also caused a curvature and pocket in space-time dragging surrounding space-time toward it.

If space-time can be shrunk then the space between Sphere/Galaxy A and B will be pulled toward and down into the pockets formed by the two Galaxies which come closer together. If space-time (the rubber sheet) were laid out straight and the two Spheres/Galaxies were suspended just above the sheet, the distance between the two galaxies would be, hypothetically, 6 feet. Because each of them have dragged space-time toward the curvatures/pockets/holes/pits they have created, they are drawn together and, hypothetically, the distance between them has shrunk to just 3 feet.

A similar phenomenon occurs if these two spheres, separated by a distance of 6 feet suspended in air, were dropped at the same time. They would fall at the same rate, but they would angle closer to each other as they fall, coming closer together, pulled toward Earth's center of gravity. And when they strike they would sink into the soil to varying degrees.

Consider just this solar system, with space-time again represented as a thin flat sheet of rubber stretched out and extending to the Oort cloud well beyond Pluto. If we place the Sun at the center, the rubber sheet will sag and dip due to the Sun's weight, forming a concave ring which can be likened to a roulette wheel. Everything which goes around the sun, including Jupiter, Mars, Earth, our Moon, is drawn toward the center where the Sun is located.

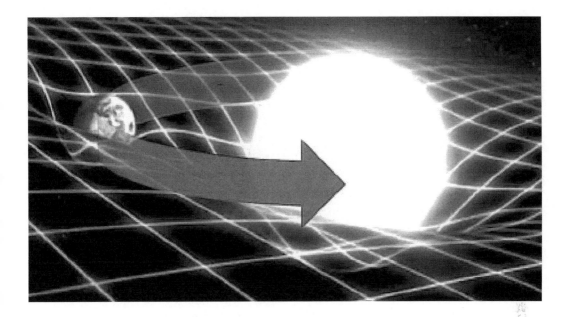

If the Sun was the only weight on this sheet, then any planets rolled along the sheet would not be able to go in a straight line but would be drawn toward the central pocket in space-time formed by the sun. The sun effects the geometry of the rubber sheet and the geometry of space-time, and like a roulette wheel, a planet would also move in a curve toward the pit. The sun curves space-time. Planets orbit the sun because they move in a path which has been carved out by the sun's gravity which forms a geometric vortex in the fabric of space-time. Its the sun's gravity which keeps the planet's in orbit, otherwise, due to their velocity and momentum, the planets would fly off into space (Carroll 2004).

A roulette wheel has numerous indentations and slots (representing different numbers and colors), and these slots could be likened to the depressions in space-time created by various planets in orbit around the Sun; Mercury forming a small concave slot, Venus and Earth forming larger concave indentations, Jupiter and Saturn forming much larger concave depressions. Each of the planets, according to weight, gravity, mass, are pressing down upon this sheet of space-time. All these circular pockets, with the largest pocket created at the center by the Sun, drag surrounding space time toward them, creating circular grooves and vortexes. The orbit of each of these planets is effected by the distortions, grooves, vortexes created by the other planets with the pocket at the center created by the Sun having the greatest effect on their orbital trajectory.

In contrast to Newton who would say that Earth moves in its orbit because of the Sun's gravitational "force of attraction," Einstein would say its a consequence of the Sun's mass which curves space-time which causes the planets to move along the curvatures, likes little roulette balls rolling along the grooves in the fabric of a space-time roulette wheel. As summed up by John Wheeler (2010), "Spacetime tells matter how to move; matter tells space-time how to curve."

Therefore, not just the space surrounding each of the planets and the sun, but the space between the planets and the sun come to be warped. Thus empty space (vacuums) can be warped by the gravity of moons, planets, stars, and galaxies.

The curvature of empty space adjacent to massive objects is predicted by

Relativity, Space Time...

so called "Schwarzschild solutions," and the relation is specified by Einstein's field equations.

Gravity Bends Light

Gravity alters the path not just of planets, but of light. Gravity bends light by altering the curvature of space-time. Light can only move in a horizontal line in a flat universe. In a curved universe and due to gravity, time and light travel in pathways described as "geodesics" (semi-circles).

Light and time are not synonymous. And yet, since time may also have energy or mass or a particle wave duality (since time is experienced) then time may also be bent by gravity. Thus, light, time, and space-time are effected by gravity, and like space-time, the trajectory of light (and thus time) can be warped and curved.

If a man and his cell phone fell off the roof of a very tall building, all would fall at the same rate, horizontal and parallel to each other, and would strike the ground at the same time. They would fall at a rate of 32 feet per second squared. This is evidence of Earth's gravity.

As conceived by Einstein, a man in an elevator which is falling because the cable has snapped, will "float" in air, above the floor, and will stay parallel to the walls of the elevator, and so will his cell phone if he drops it. The cell phone and the man will "float" in air horizontal and parallel to each other above the floor so long as the elevator falls at a constant velocity. If the elevator has a wall light which is emitting a beam of light, the light beam will strike the opposite wall in a horizontal straight line. Thus, the man, the cell phone and the beam of light would remain parallel and horizontal to each other even as they fall to their doom.

However, if the falling elevator were to suddenly accelerate upward, at a rate of 32 feet per second which is equal to 1 g of gravity, the gravity of Earth, the man and his cell phone will fall to the floor at a rate of 32 feet per second squared. They will strike the floor as if the elevator were sitting on the Earth and not accelerating when in fact the floor is rushing up and smacking into them. However, the light beam will not strike the opposite wall in a horizontal straight line. Instead, the beam of light will bend at an angle and will strike at a lower portion of the opposite elevator wall. As the elevator accelerates upward it causes light to bend and curve downward.

Acceleration creates g-forces. Acceleration involves energy which can become mass. This can be predicted from Einstein's formula, $E = mc^2$. Light is a form of energy and energy is equal to mass squared, and mass is effected by gravity and has gravity. Hence, it would appear that light may also have mass (represented by its particle, the photon) and gravity; though many physicists would argue with these latter conclusions.

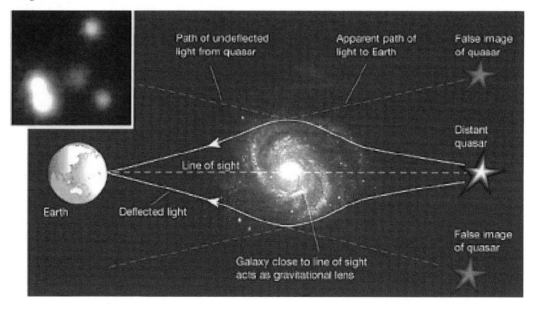

Figure: Gravitational Lensing. Gravity of galaxy splitting and bending light

Photons (particles of light) are said to be massless when they are at rest. However, light has energy and momentum, exerts pressure on various substances, and can be bent, but is never at 'rest." Thus, photons have potential mass and this is because a photon has energy $E = hf$ where h is Planck's constant and f is the frequency of the photon. And since a photon has momentum, and momentum p is related to mass m by $p = mv$, then a photon has potential mass. And since photons have energy and energy is equivalent to mass according to Einstein's formula $E = mc^2$ then a photon has mass.

Gravity can also warp and alter electromagnetic activity; causing charges to accelerate and creating a force, or energy field which can effect distant locations. Differential gravity and electromagnetic activity also creates an imbalance which can lead to electromagnetic diffusions and the conductance of electrical charges. Electromagnetism and gravity are not the same thing. Whereas all objects are accelerated in the same way by gravity regardless of their mass, when subject to a magnetic or electric force, the mass and charge of the object will determine its rate and speed of acceleration, such that different "charge-to-mass" ratios are accelerated at different speeds.

Just as the geometry of space-time can be curved, bent, twisted and distorted by gravity and acceleration, acceleration and gravity can curve and bend light, and that would include the gravity of the sun, Earth, this galaxy, and so on (Carroll 2004). Since time must also have energy, then as light goes, so too does time.

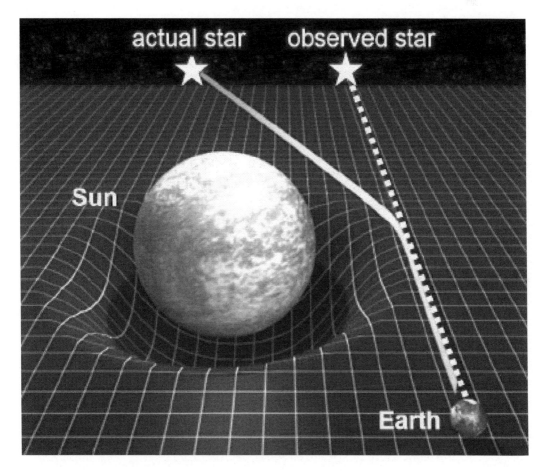

Figure: Gravitional lensing and refraction causing light of stars to bend

Quantum Physics of Time Travel

Einstein's theory on gravity and the bending of light was first confirmed in 1919 during a solar eclipse, when Arthur Eddington and Frank Watson Dyson observed that light from stars passing close to the Sun was slightly bent, so that stars appeared slightly out of position (Dyson et al. 1920). Einstein (1911, 1916) predicted a deflection of 1.75 seconds of arc. When measurements were taken from two different locations, Brazil and an Principe Island, and when pictures of stars near the sun before and after a total solar eclipse and when the sun was on opposite sides of the sky, were compared, Einstein's prediction was confirmed. However, his theory also predicts that as gravity increases space is increasingly bent, which dramatically alters the path of light and which can also punch holes in the tissues of space-time.

Figure: Refraction. Light can be bent by its surroundings

Relativity, Space Time...

Galactic Lensing

Consider again Spheres/Galaxies A and B and the depressions/pockets created in space-time. Because of this curvature light from Galaxy B would curve around Galaxy A (and vice versa) thereby becoming distorted and bent. This also implies that time may become distorted and bent.

Light can be split apart, and the same is true of the light of galaxies which pass other galaxies, a splitting which creates multiple images as the light splits and coils around these galaxies; a phenomenon known as "Gravitational lensing" (Renn et al., 1997; van der Wel et al. 2013). If time can also be split apart, is unknown; though it could account for the experience of what is called the past vs the present vs the future. If light carries images from the past of a distant star, and thus images from the past, then, perhaps the past (or rather, that past) can also be warped and bent. Since that light, before it is received on Earth is in the future relative to Earth then perhaps the future can also be bent, warped and split apart.

Quantum Physics of Time Travel

The sun is just one star among of up to 400 billion other stars within the MilkyWay galaxy (Villard 2012). The Andromeda Galaxy is believed to contain over a billion stars (Young 2006). Each of these billion stars are no doubt orbited by planets. These stars and their planets will create pockets in and warp surrounding space-time which is dragged toward them. Collectively a single galaxy might contains trillions of pits and pockets in surrounding space-time, all of which effect adjacent space and the trajectory of light.

For example, a cluster of galaxies called SDSS J1004+4112, which is 7 billion light years from Earth, lies directly in front of a more distant galaxy which is 11 billion light years from Earth (Oguri 2010). Light from that more distant galaxy is bent by the tremendous gravity and numerous holes and pockets created in space-time by galaxy cluster SDSS J1004+4112, such that multiple images of this more distant galaxy appear alongside this cluster, when in fact it is 4 billion light years directly behind it.

The splitting, curvature, and creation of multiple images are due to the tremendous gravity and multiple curvatures and pits creates in space-time by the trillions of stars and planets which make up these galaxies, each of which causes the light from this more distant galaxy to be split and sent spinning in different directions. Instead of a single ball going round and round a roulette wheel, the ball is split into multiple balls.

Space-time, therefore, has numerous curvatures and pits and pockets of varying size, all of which effects the trajectory of light (Carroll 2004). Since time

Relativity, Space Time...

is part of the 4th dimension of space-time and is related to light, then time is also effected, and like light, time may also be split, coiled, curved, and bent. Instead of a single uniform "universal time" there are innumerable pasts and futures and infinite "nows" some of which may arrive side by side, or one after the other, or even curl around such that the future leads to the past.

8: The Circle of Time: In A Rotating Universe The Future Leads to the Past

Einstein (1915a,b, 1961) theorized that time and space can be unified in the 4th dimension. Like the unification of mass and energy, space-time are two aspects of the same quantity, such that space can be converted into time, and time into space in the 4th dimension. Space-time and time, therefore, have energy, and can be experienced and perceived, and in this respect, time also shares characteristics with light and may have a particle wave duality.

A fundamental principle of physics is that a beam of light takes the shortest path between two points which is a straight line (Fermat's least time principle). However, light bends due to the influence of gravity (Einstein 1911), which means the path is not straight, but curved. Likewise, according to Einstein (1914, 1915a), space is curved. However, that curvature would not be a round circle, but would have different geometric characteristics depending on and due to the unequal influences of gravity in various regions of the cosmos.

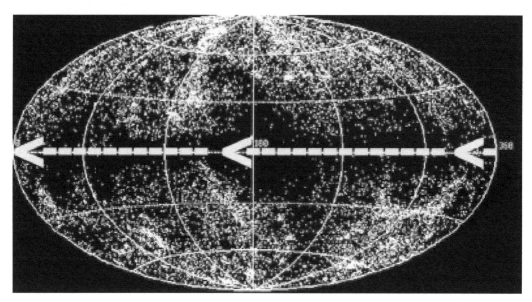

Figure: Curved Hubble Length Universe Depicting Clustering of Galaxies: Tavel in one direction leads back to the starting point.

If the entire universe is curved, as predicted by Einstein's theories, then just as traveling in a straight line on Earth will bring the traveler full circle to his

Relativity, Space Time...

starting point, the same could be applied to a curved universe as well as to the trajectory of light and time. Time, like time-space, has curvature; and just as a journey in a "straight" line will bring a voyager full circle around the globe, the same could be said of a journey across space-time. Time may be a circle; a cosmic clock which ticks at different speeds depending on gravity and the geometry of space-time relative to an observer. However, what this also implies is that a journey across time will bring the voyager full circle, such that the present leads to the future, and the future leads to the past.

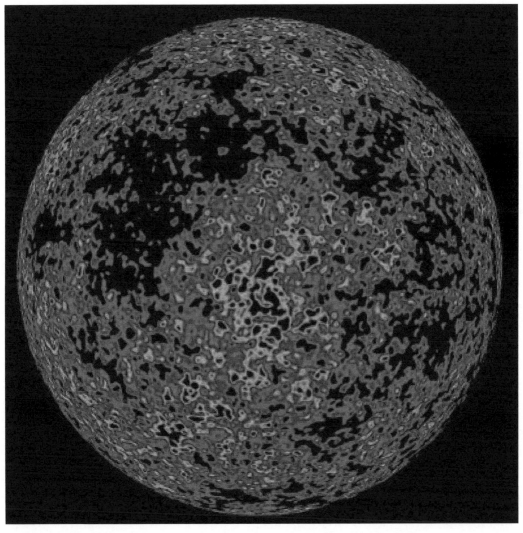

Figure: Wilkinson Microwave Anisotropy Probe (WMAP) map depicting a curved "Hubble Length Universe" and galactic "voids" billions of light years in size.

Quantum Physics of Time Travel

Because gravitational influences vary throughout the cosmos, then every infinitesimal region of space-time may have its own proper time relative to observers in different locations. The present on a distant galaxy, as conveyed by images of time-light, does not arrive on Earth until the future, such that the future and the past overlap in time-space. A logical corollary is that there is no universal "now" past or present, and that "absolute time" does not exist (Gödel 1949a,b)

Beginning in 1949, Kurt Gödel, in a series of papers based on Einstein's field equations of gravity, rejected the Newtonian conception of time and prevailing beliefs that the "present" consists of infinite layers of 'now' coming into existence in continual successive and immediate sequences. According to Gödel (1949a,b, 1995) if space–time is curved, then the experience of time could be considered a consequence of that curvature.

As based on Einstein's field equations, Gödel (1949a,b) discovered that a particle traveling through space would circle round from the present to the future and then continue to circle around and meet itself in the past; and from the past that particle would circle round and meet itself in the future and from the future it would again travel round and meet itself in the past; an infinitely repeating pattern.

Gödel argued, since space-time is curved, then the future and past may also be curved and circle round thereby completing the circle which then continues in an endless loop. Time is a circle.

According to Godel, because of the curvature of time and space it is possible to travel through time: "By making a round trip on a rocket ship in a sufficiently wide curve, it is possible in these worlds to travel into any region of the past, present, and future and back again."

Gödel's formulations also borrowed from George Gamow's (1946) conception of a universe which, like all astral bodies in space-time, is in orbital motion. As pointed out by Gamow and Pythagoras 2000 years before, patterns repeat themselves in nature from the subatomic to entire galaxies (Joseph 2010c). Electrons orbit the nucleus of the atom. Planets orbit the sun. The sun is just one of billions of stars located throughout the spiral arms of the Milky Way Galaxy, and the entire galaxy is rotating. Perhaps the entire universe is also rotating.

Closed Time Curves In A Rotating Universe: The Future Leads to the Past

Gödel (1949a,b) explained that if the universe was rotating and space-time is curved, time should also be curved and curve back upon itself, forming infinitely repeating closed time-like curves (CTCs). Just as it is possible to circle the Earth and return to where one began, if time and time-space circles back on itself in Pythagorean endless loops thereby giving rise to CTCs, it would be possible to journey in a circle back to where one began; which means, one can travel into their own past.

Any journey along a CTC eventually leads the time traveler back to the point in space and time from where he started, even to the day he was born. In a "rotating universe" time is a circle where the future leads to the past and effects precede causes; the future can effect the present, and the past. However, the time-traveler journeying along such a loop does not experience a slowing of time as there is no contraction of space-time. Time would remain the same for the time-traveler and all those on Earth, as the time traveler is merely going in a circle.

A rotating universe and closed time-like curves violate the rules of causality. If time can circle back on itself, then the future can effect the past and the temporal discontinuity between past, present, and future is abolished. Hypothetically, since cause and effect are abolished, if, as an adult, you do something bad in the future when you become an adult, you may be punished for that future indiscretion while you are a still a child; like "karma" in reverse.

Multiple Earth's In Curved Space Time

A Gödel rotating universe also implies duality, if not multiplicity; and the same is true of the bending of light. Because gravity increases the curvature of space-time and can bend and split images of light, as illustrated by galactic lensing (Renn et al., 1997; van der Wel et al. 2013), then light reflected from Earth may also be split apart as it curves through space. When these light-images of Earth cross paths with stellar objects of sufficient gravitational strength, these beams of light may be curved round in an 180 degree arc with Earth as its target. That is, light-images, or time-light, may be split apart and circle around numerous gigantic galaxies, and some of these light images will be reflected back toward Earth and become mirror images of Earth's past. Therefore, as we gaze at the various stars which twinkle in the darkness of night, some of those *"stars"* may be images of Earth and our solar system from the long ago.

In a Godel spinning universe, the mirror would also be gazing back; meaning that this Earth could also be a mirror from the past, a reflection that those on a future Earth can look back upon.

Closed Time Light Curves: The Future Causes the Past

There are two conceptions of time-like curves, open or closed (Bonor & Steadman, 2005; Buser et al. 2013; Friedman et al. 1990) . Open time-like curves follow an arrow of time straight into the future, and there is no return to the past unless one can exceed the speed of light.

Time-like curves which are closed loop back in a circle; meaning that future events could affect past ones. A closed time-like curve (CTC) is a world line in a Lorentzian manifold, such that the particle, object or time traveler returns to its starting point (Buser et al. 2013). That is, because light is curved, then light can loop back on itself, and it would be possible for an object to move around this loop and return to the same place and time that it started. An object in such an orbit would repeatedly return to the same point in space-time.

Quantum Physics of Time Travel

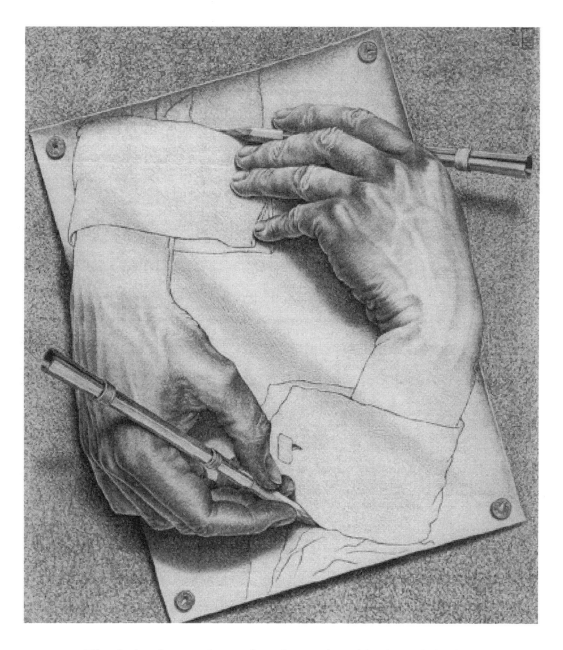

The circle of space-time points forwards and backwards in time. If CTCs exist then it would be possible to travel backwards in time. The question becomes: how far back or forward in time? For example, it may be possible to follow the CTC in a negative direction and revisit the day President Abraham Lincoln was assassinated, the morning when Jesus was nailed to a cross, 65 million years ago when a giant asteroid struck this planet eradicating the last of the dinosaurs from Earth, and further back still to the Cambrian Explosion 500 million years ago when all manner of complex species with bones and brains appeared almost simultaneously in every ocean of Earth, and to the time when Earth was hellishly hot and populated by only microbes some 4.2 billion years ago.

Relativity, Space Time...

Einstein's theories of relativity (despite his posting of a cosmic speed limit) predicts that the only way to travel into the past is to exceed the speed of light. Upon accelerating toward light speed, space-time contracts and the space-time traveler is propelled into the future. However, it is only upon accelerating into the future and then beyond light speed that the contraction of space-time continues in a negative direction and time flows in reverse. It is only at superluminal speeds that time reverses and one can voyage backward in time. Einstein's general theory of relativity predicts that the future leads to the past. Likewise, as shown by Gödel 1949a,b), Einstein's field equations predict that time is a circle; and this violates the laws of causality (Buser et al. 2013).

If time is a circle, then effects cannot always be traced to an earlier cause, for the cause may occur in the future. In a closed loop, the future can come before the past and catch up with itself in the past so that an an event can be "simultaneous" with or occur before its cause. An event may be able to cause itself.

These are not just thought experiments. There is considerable evidence of what Einstein (1955) called "spooky action at a distance" and faster than light "entanglement" (Plenio 2007; Juan et al. 2013; Francis 2012). It is well established that causes and effects can occur simultaneously and ever faster than light speed (Lee et al. 2011; Matson 2012; Olaf et al. 2003); a consequence of the connectedness of all things in the quantum continuum.

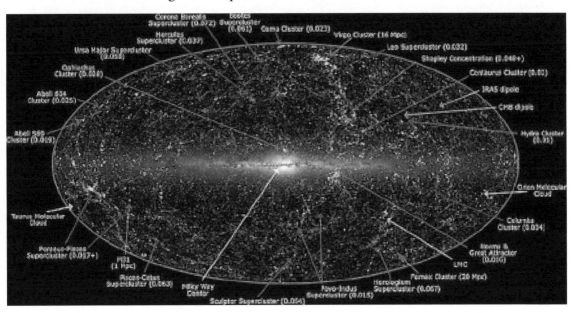

Figure: Clustering of galaxies in the "Hubble Length" visible portion of the universe.

Einstein's curved universe could not be a perfect circle, as galaxy distribution is asymmetry, with great "walls" of galaxies throughout the cosmos which have clustered together. It is this clustering, and these galactic walls which contribute to the unequal distribution of gravity, which causes space-time not just to curve, but to fold and curl up and to asymmetrically effect the flow of time.

Gravity is always strongest at the center of gravity where its most concentrated. Time-space is also be pulled toward the center of gravity, which is why Einstein proposed his "Cosmological constant" a repulsive force which would prevent the universe from collapsing. Einstein later rejected his "cosmological constant" calling it "the biggest blunder" of his life, when in 1929 Hubble reported the universe was expanding. Einstein believed that if not for his "cosmological constant" he could have predicted an expanding universe. Instead, the prediction was made by Alexander Friedmann in papers published in 1922 and 1924.

However, as pointed out by Godel (1949a,b), Einstein's equations do not predict an expanding universe, but a rotating universe; a conclusion that Gamov (1946) also arrived at years before based in part of his observations of rotational patterns throughout the cosmos.

Patterns, Symmetry, Rotating Universes and the Circle of Time

A pattern, be it recurring numbers, events, or objects, repeats itself in a predictable manner down to its essential elements (Ball 2009; Novak 2002; Wille 2010). The entire field of mathematics is the "Science of Patterns" and any sequence of numbers that may be described by a mathematical function has a pattern (Wille 2010). The pattern at the elementary level is the basis, model, or template which is repeated on a larger scale to generate larger objects or series of events all of which exhibit the same or similar underlying pattern. Hence, elementary particles have orbits, planets have orbits, stars have orbits, and it can be assumed that, collectively, galaxies have orbits which would mean the "known" Hubble length universe, is also in a rotational orbit as all share similar patterns (Joseph 2010b).

Figures: Sea shell. Snail Shell, Pythagorean "Golden Ratio".

In Euclidean geometry, a pattern known as a translation involves movement of every point at a constant distance in a specified direction and the same can be said of rotation (Johnson 2007). The symmetry of the cosmos is based on the repetition of patterns found throughout nature, from sea shells to spiral galaxies (Joseph 2010b). For example, snail shells, sea shells, vortices, the cochlear nucleus of the inner ear, etc., show similar repeating patterns around an axial center or "eye." The patterns intrinsic to the shell of a snail are replicated repeatedly in nature and typify the structure of whirlpools, cyclones, hurricanes, the Milky Way galaxy and every spiral galaxy so far observed all of which rotate around an axial point (a "black hole") at their center.

Since rotating patterns repeat, as pointed out by Gamov (1946) and 2000 years earlier by Pythagoras, then the universe (including an expanding universe) would be part of this pattern. The entire universe, therefore, must orbit and rotate around an axial point, as predicted by Einstein's field equations (Gödel 1949a,b). Again, however, Einstein's equations do not predict a perfectly curved universe, but a lumpy universe with waves and crests which circles round and which would be pulled inward toward the center of gravity; that is, if the universe is considered as a collective single entity. If correct, then curvature of space-time would continue as a repeating pattern of curvature, curving forever inward and outward; which leads to Pythagoras and the "golden ratio" (Joseph 2010b).

Figure: Hurricane Francis off the coast of Florida

Because of gravity, time-space can also be bent backwards in a circle, as happens with whirlpools and eddies along river banks where water flows in a circular motion. If the implications of Einstein's theories are correct, then the river of time is bent round in a circle and it has no ending or beginning and may include pockets or vortexes of time which pop in and out of existences like vortexes and eddies along the river banks.

Quantum Physics of Time Travel

Einstein's time-space curvature coupled with gravity and the principle of repeating patterns, and the concept of close-time-like curves, raises the possibility that space-time is like a spiral staircase, so that when circling round at 360 degrees one does not end up in the same space or spot where they began, but above or below it (Buser et al. 2013; Gödel 1995). From the perspective of the Time Traveler, this spiraling circle can lead to the future or the past, or to multiple futures and pasts which may coexist, in parallel, side by side.

CTCs open the possibility of a world-line which is not connected to earlier times in this past, but to multiple possible pasts, and futures, which exist in parallel, above, or below, or alongside one another--multiple spiraling staircases of time which lead to parallel worlds of time (Buser et al. 2013; Gödel 1949a,b).

Black Holes, Red Shifts, and Cosmic Evidence For a Rotating Universe

Gödel (1949a,b) developed Gamow's (1946) concept of rotating universes as a thought experiment and as a logical extension of Einstein's field equations of gravity. However, the implications were so profound, and so contrary to the predictions of Newtonian physics, Einstein's concept of relativity, and what is now referred to as the "Standard Model" that the possibility of a rotating universe has been almost universally rejected (Buser et al. 2013). Even Gödel (1949a,b, 1995) who published his observations in the 1940s and 1950s, pointed out that there was a yet no evidence of red shifts in the distant regions of the cosmos which would support a rotation model.

Gamow (1946) who first proposed a model of a rotating universe blamed the lack of evidence on the insufficient power of the telescopes available to astronomers and physicists at that time and proposed that proof of rotating universes would have to wait until advanced telescopes became available.

As based on the observation of planets, stars, and the rotation and combined gravity of mass aggregations such as entire galaxies, Gamow (1946) thought it was only logical that the entire universe must also be rotating around some axial point in space. As pointed out by Gamow, "galaxies are found in the state of more or less rapid axial rotation" contrary to the Big Bang theory and in contradiction to the belief that galaxies formed following the condensation and angular momentum of the primordial matter. Gamow posed this question: since planets, stars, and galaxies are rotating then perhaps "all matter in the visible universe is in a state of general rotation around some centre located far beyond the reach of our telescopes?" In the 1998, observations published by two separate teams inadvertently provided that evidence, as based on the red shifts of distant stars which had undergone supernova (Perlmutter et al., 1998; Schmidt, et al., 1998).

Gamow (1946) based his rotating universes model on the rotation and angular momentum of galaxies which appear to orbit an axial point in space. Any rotating body, be it a galaxy, a merry-go-round, or the planets orbiting the sun, shows differential speeds of acceleration and velocity depending on how far away they are from the axial center of rotation. For example, in the inner galaxy, the rotation speed rises with the radius. By contrast, in the outer galaxy the rotation speed remains constant (Petrovskaya, 1994; Teerikorpi, 1989). The point closest to the axis rotates faster than points closer to the outer rim which rotate at a similar velocity. For example, Earth and our solar system, located on an outer arm of the Milky Way galaxy, orbit the supermassive black hole at the axial center of the galaxy, at a speed of approximately 155 miles/sec (250 km/sec) (or from 965,600 km/h, to 804,672 km/h), taking around 240 million years to complete an orbit. However, those stars closest to the axial galactic center, relative to the stars on the outer rims, are moving more rapidly and display accelerating velocities as they come closer to the central axis (Ghez et al., 2005; Petrovskaya, 1994; Teerikorpi, 1989). In fact, the speeds are so high they are beyond what would be predicted based on the universal law of gravitation (Schneider, 2006); observations which also led Gamow (1946) to question the Big Bang origins model and to propose that the universe may be in rotation.

Supermassive black holes are believed to lie in the center of every spiral galaxy, thereby dragging time-space into the hole, and creating massive geometric vortexes in space-time around which orbits all the stars of the galaxy. In addition to their incredible gravity which can be equal to a billion suns, Black holes have spin (Nemiroff, 1993) thereby creating a vortex in the geometry of space time such that stars closest to the black hole have a greater orbital velocity compared to those further away (Ghez et al., 2005; Petrovskaya, 1994; Teerikorpi, 1989), with stars and space-time eventually being dragged down into the hole. Black holes therefore warp the gravity of time-space and are believed to exert an organizational and gravitational effect on matter and attract and organize stars which then circle and orbit around them, much in the same manner that water is drawn toward and then circles 'round a drain before disappearing inside.

Those stars nearest to the axial black hole center of the galaxy display a characteristic red-shift pattern of emitted light which is indicative of their acceleration toward the hole and their great distance from Earth (McClintock, 2004; Melia 2007; Thorne & Hawking 1995). The concept and interpretation of redshifts are based on the Doppler effect.

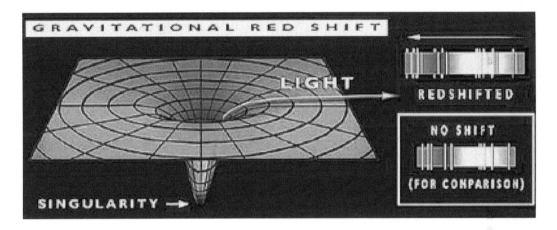

For example, the sound of an approaching siren changes frequency as it approaches and then passes a stationary observer. Likewise, the frequency of light waves also change, such that upon approach the frequency of its waves increases, and the light becomes increasing blue in color; that is, it displays a blue shift as the waves crowd together. However, once the light wave passes by an observer the frequency decreases and the waves grow further apart and increasing red in appearance as it recedes away into the distance. Therefore, if a source of the light is moving away from an observer, then redshift ($z > 0$) occurs. If the source moves towards the observer, then blueshift ($z < 0$) occurs. This is true for all electromagnetic waves and is explained by the Doppler effect. Stars closest to the black hole, and which are accelerating toward their doom, display this red shift.

If the universe is rotating, then stars nearest the center of the universe should be rotating at a faster velocity, and also display a red shift indicative of acceleration and their great distance, relative to observers further from the central axis of rotation. That is, like those stars in the inner axis of the galaxy which have a greater velocity and are furthest from Earth and which display these red shifts, the same should be true if this universe is in rotation; those stars closest to the axis should display a red shift indicating they are moving faster than stars further from the axis.

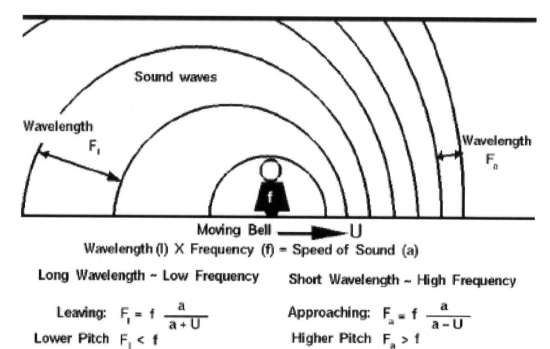

In 1998 two teams of scientists discovered that distant stars over 8 billion light years distant from Earth, and which had undergone type IA supernova (SN IA) were displaying red-shifts which were more extreme that stars closer to Earth (Kirshner 1999; Perlmutter et al., 1998; Schmidt, et al., 1998). Thus, these distant SN IA stars were not only moving away but were accelerating. As detailed by Robert Kirshner (1999), a member of one of the teams which made the acceleration discovery: "...the distant supernovae are explosions that took place 8 billion years ago....The distant supernovae are not brighter than expected in a coasting universe, they are dimmer. For this to happen, the universe must be accelerating while the light from the supernova is in transit to our observatories.... Could there be some other reason... that makes the objects found at a redshift $z = 0.5$ approximately 25% fainter than the SN Ia we see nearby?"

The teams led by Perlmutter and Schmidt interpreted their results based on a Big Bang model of the universe, and therefore as proving the universe was expanding and the expansion was accelerating. The discovery of acceleration was a surprise and cannot be explained by any of the competing theories based on a Big Bang beginning for the universe, other than by appealing to unknown forces such as "dark energy" and "dark matter."

An alternate explanation is that the red shifts associated with these type IA supernova (SN IA) are evidence of a rotating universe (Joseph 2010b,c) as predicted by Einstein's field equations and as proposed by Godel and Gamow. If the universe is rotating this would explain the presumed increases in the velocity of distant stars and galaxies: they are repeating a spiral pattern found throughout nature. The most distant stars and galaxies may be rotating at a more rapid velocity as they are closest to the axial center.

Time is a Circle

Einstein's field equations predict a universe which curves back upon itself, such that anyone traveling in a wide enough circle across the cosmos would return to the point at which they began their journey. Although the present may lead to the future, the future leads to the past.

And if light follows a closed time-like curve, and since time, like light, may be curved, split, and turned around, then time-light images of Earth from the long ago may be mirrored throughout the cosmos, along with images of innumerable other stars and planets from the long ago.

There may be multiple reflections from Earth's past, providing windows into events which took place 4.2 billion years ago when the first evidence of biological activity appeared in this planet's oldest rocks, or 60 million years ago when a comet or asteroid slammed into this planet giving the dinosaurs a coup de grace, or to November 22, 1963 providing us with a bird's eye view of Dallas, Dealey Plaza, the grassy knoll, the book depository building and the assassination of President John F. Kennedy, resolving once and for all one of the greatest "who done its" of the 20th century.

Time is a circle, and this too is predicted by Einstein's theories of relativity, where accelerating toward light speed takes the voyager to the future, and upon exceeding the speed of light, the traveler heads back from the future into the past. However, upon reaching light speed, not just space-time, but the time traveler may shrink so much in size and thus have so much mass and energy, that the time machine punches a hole in space-time; a hole through which the time traveler can quickly journey to a mirror universe on the other side.

9: Time Travel Through Black Holes in the Fabric of Space-Time

If you drop two balls at the same time from the roof of a building, one weighing 10 pounds and the other weighting 100, they will fall at the same rate and hit the ground at the same time due to gravity. However, the heavier ball will form a larger crater in the soil and this is because of their differential mass and gravity. The same can be said of planets and galaxies in space-time.

Gravitational fields are not uniform and differ in strength in various regions of this galaxy (Carroll 2004); just as the weather is different in distant localities on Earth. Galaxy distribution is asymmetric with great walls of galaxies clustering together. The planets and stars orbiting these galaxies, the galaxies themselves, and the clustering of these galaxies all differentially effect and torque the geometry of space-time such that it is littered with pockets of varying size.

Relativity, Space Time...

The effects of gravity differ throughout the cosmos and even on the same planet. On Earth, gravity is stronger at sea level than at higher elevations, and even stronger toward the center of the planet. Hence, objects are pulled down toward the center of gravity.

All falling objects are drawn toward the center of Earth's gravitational attraction--the center of Earth. If two identical objects are placed hundreds of miles apart miles above Earth and allowed to fall, they will come closer and closer together as they fall in a pattern similar to an inverted triangle, until they end up side by side and pointing directly toward Earth's center. Because of Earth's curvature, the balls are pulled together, toward the center of gravity which is at the center of Earth.

If the two balls are placed directly above the center of Earth, but with one ball 50 feet directly above the other, then as they fall, the lower ball will fall at a slightly faster rate than the upper ball, because it is closer to Earth's center and experiences a stronger gravitational pull.

Earth is comprised of layers. The outermost layer is referred to as the crust, beneath which is the mantle which is made up of heated rock under high pressure, and below that is believed to be exceedingly hot liquid metal (with an estimated temperature of around 7,000 kelvin) and a compressed metal core.

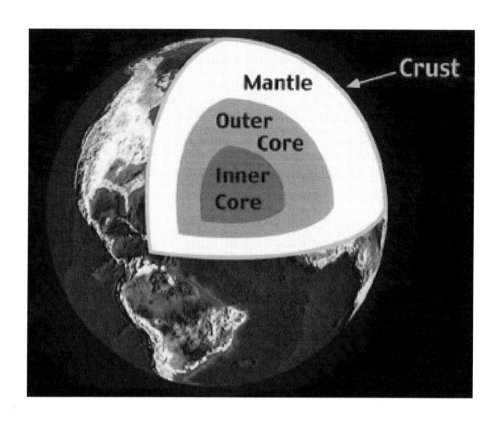

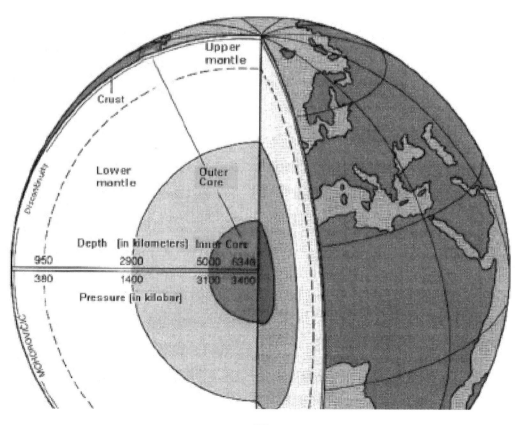

Relativity, Space Time...

Earth has a mass of 5.9736 x 10^{24} kg (5973600000000000000000000 kg), an equatorial diameter of 12,756.1 km and radius of 6,378 km. If the inner layers of Earth were to collapse and the planet imploded to half it size, the outer surface would be pulled toward the Earth's center, the planet's gravity would double, and time would slow. All the clocks and survivors on Earth's surface would also move very slowly from the perspective of those aboard the International Space Station (ISS) in orbit above the shrunken planet. However, if astronomers on Earth were looking back at the crew of the ISS, then everything taking place within the ISS would seem to have speeded up; a predicted by Einstein's (1961) theories of relativity.

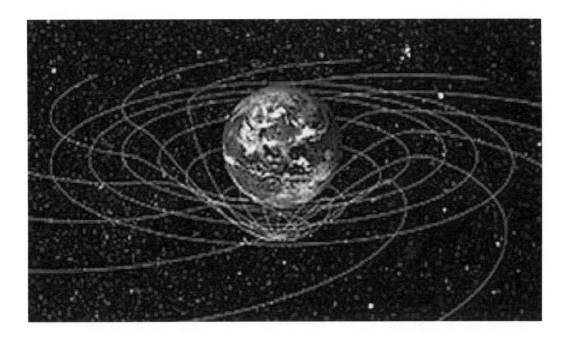

The gravity and mass of Earth deforms and creates a deep pocket in the geometry of surrounding space-time (Carroll 2004). If Earth were to implode, the increased gravity would create a deeper pocket, or pit in space-time and drag surrounding space time toward it. If Earth continued to implode, thereby concentrating its mass, Earth's center of gravity would increase it's gravitational power, pulling more of the outer surface toward it. Those on the surface would begin to shrink and sink into the ground and their movements (and time) would be so slow they would appear to be frozen in place. If its radius and diameter continued to shrink the collapse and implosion of Earth would accelerate due to the increasingly powerful pull of gravity toward its center. Soon what had been a planet with a radius of 6400 km and a mass of 5.9736 x 10^{24} kg, would become the size of a golf ball with a radius of just a few centimeters while retaining the mass and gravity of an entire planet.

Relativity, Space Time...

Moreover, because of the energy involved in its acceleration toward miniaturization, it will increase in mass and gravity even as it grows smaller becoming molecular in size. Einstein's and Newton's theories of gravity both predict that if mass is shrunk to a subatomic space, its gravity will become increasingly powerful (Einstein 1915a,b, 1961) thereby punching a deeper hole and creating a powerful vortex in space-time. In consequence, anything on the surface would be crushed to atomic size, and passing light would be sucked down into the hole

The Earth would continue to shrink to the size of an atom and would no longer be visible. No reflected light would be able to escape, and time would stop due to the forces of gravity which would prevent the movement of time. Instead there would be a tiny black hole in the fabric of space time; and this atom sized planet would be at the center, halfway between the top and the bottom of the hole. Passing light would continue to be pulled down into this hole and to the surface of this miniature planet. Because no light can escape, the hole would be black.

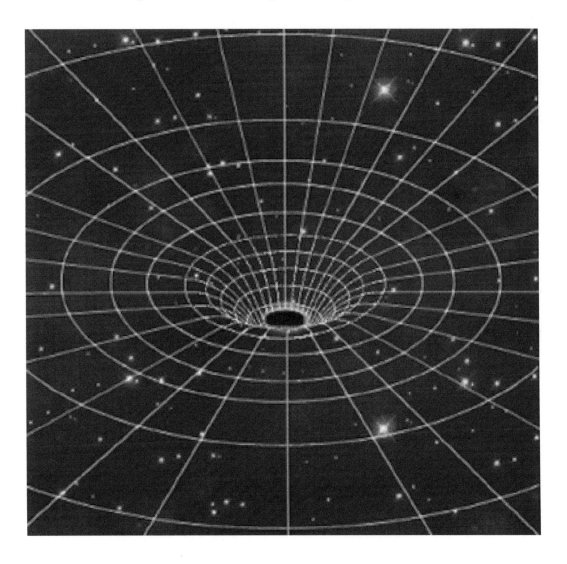

Quantum Physics of Time Travel

However, if the shrinkage continued, and the Earth became smaller than an atom and less than 10^{-13}cm in size, then, as summed up by Heisenberg (1958), the result is "time reversal...The phenomenon of time reversal...belongs to these smallest regions." Time travels in reverse in these tiny spaces because of the tremendous energy released which blows a hole through the tissues of space-time which tunnels from the present to the past at superluminal speeds. Gravity is so powerful in spaces smaller than 10^{-13}cm, that it can suck time backwards at such incredible speeds that time itself would exceed the speed of light thereby propelling everything in its wake into the past.

Holes in space-time offer a passageway, what has been referred to as an "Einstein Rosen bridge," which leads from the present to the future, and to the past.

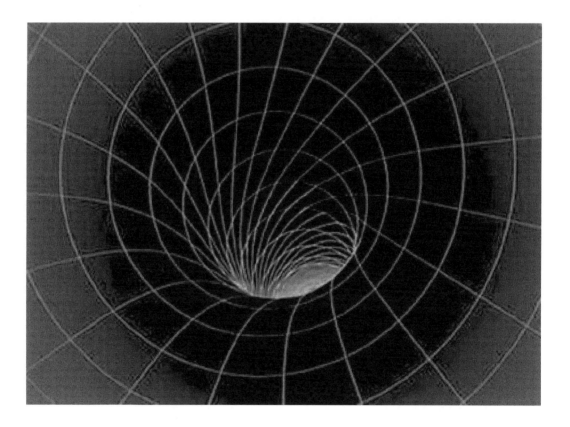

The cosmos is believed to be littered with holes of varying size and magnitude, from those smaller than an atom to super-massive black holes with the gravity and mass of entire galaxies (Al-Khalili 2011; Hawking 1988; Joseph 2010c Thorne 1994).

If a time traveler wished to journey into the past he could shrink himself in size to smaller than 10^{-13}cm (impossible with current technology) or dive his time machine into a supermassive black hole--because according to Einstein and

Relativity, Space Time...

his colleague Nathan Rosen, the hole has no bottom and tunnels to a mirror universe on the other side (Einstein & Rosen 1935).

Tunneling Through Time

Just as a planet is curved into the geometry of a circle, space-time is curved and, theoretically, this can allow for "short cuts" between planets, solar systems, and entire galaxies. For example, China and Argentina are antipodal, on opposite sides of the planet. The distance in a curving "straight line" between between Beijing China and Buenos Aires Argentina, is 12,326 miles.

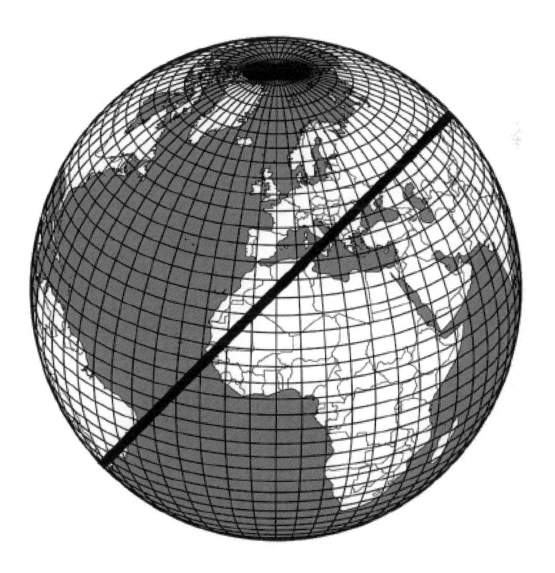

Quantum Physics of Time Travel

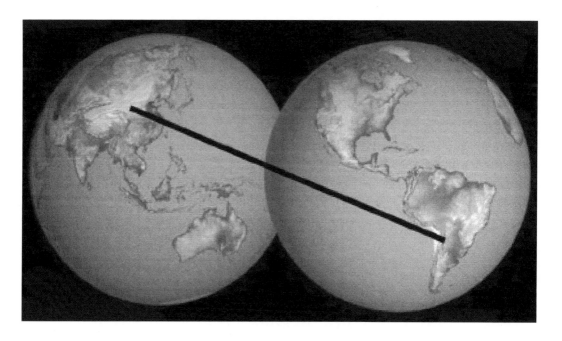

If instead, one could travel directly from Argentina to China, by tunneling from the surface through the center of Earth to the other side, the distance would only be about 7,900 miles, a reduction by 3,426 miles (35%). Theoretically, just as one might drill a hole tunneling downward from Argentina, and end up on the other side of the planet in China, a hole in the curvature of space-time may also lead to a galaxy or a universe on the other side.

If a sufficiently massive object such as a super-sized star or galaxy were to collapse and implode, they may not simply sink deeper into the original pocket or pit which their gravity had already carved into the tissues of space-time, but they may punch a hole that has no bottom but opens at the other end, creating a passageway; or what has been referred to as an Einstein-Rosen Bridge, to the other side (Einstein & Rosen 1935).

Relativity, Space Time...

Holes forming an Einstein-Rosen Bridge have such immense gravity that not just space-time, but light is sucked down into the hole (Bethe et al., 2003; Melia 2007; McClintock 2004). Although Einstein's theory of special relativity erected a cosmic speed limit which proclaimed nothing can exceed the speed of light (1905b, 1906a), his theory of general relativity, of gravity, abolished the speed limit (1915a,b). Under the influence of tremendous forces of gravity, the speed of light can be exceeded; and this is what happens to light sucked into the mouth of a supermassive black hole (Melia 2007; McClintock 2004; Thorne & Hawking 1995). And when light, or any object exceeds light speed, they are flung into a mirror universe where time runs backwards into the past.

Accelerating toward light speed collapses and shrinks space-time, compacting more time into a smaller space; shrinking the distance between the present and the future (Einstein et al. 1961). At light speed, time stops. However, once the cosmic speed limit is surpassed, the compression of space-time implodes and turns inside out continuing in a negative direction. Time and space-time are reversed, like looking into a mirror, except that the time traveler has entered the mirror and is looking back. However, the only way to reach the mirror universe is to speed faster than light and into the future, and/or to shrink to a size smaller than 10^{-13}cm. Those who accelerate beyond the speed of light journey into the future and then through the looking glass into a mirror universe which leads from the future backwards in time.

The mirror universe is not fancy, but a mathematical fact based on the Schwarzschild solution of Einstein's equations; commonly used to calculate the gravitational field of a massive star (Einstein 1915a). The collapse of supermassive stars creates super massive black holes in space-time.

Stars are born and they die, and the larger stars have a spectacular death, literally going out with a big bang, a supernova explosion, at which point they begin to contract. Those which are three times the size of our sun, are believed to collapse into black holes. Smaller stars collapse into what are referred to as neutron stars (Becker 2009).

Neutron stars are the remnants of collapsed stars similar to or a few times larger than the mass of the sun (Becker 2009). Consider for example the "Crab Nebula" deep in the constellation of Taurus which exploded in a vast supernova in 1054 and then collapsed. Gravity is so powerful that the atoms which made up this star have been crushed into neutrons and light is unable to escape its surface, meaning that time has stopped.

Stars collapse after they burn up their internal hydrogen and helium fuel causing them to expand and becoming red giant as they eject mass into space; and then they begin to implode with the increasing concentration of mass and gravity exacerbating and accelerating the implosion until shrinking and becoming compressed to a singularity, perhaps as small as a single atom, and with a density of about 5×10^{93} grams per cubic centimeter (Bethe et al. 2003; McClintock, 2004).

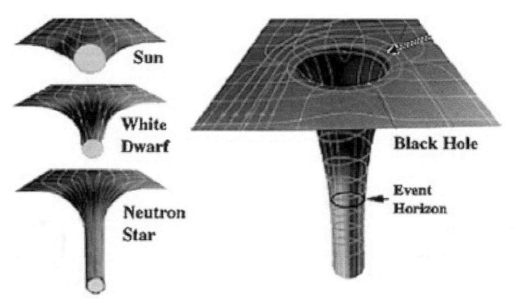

Because of its incredible concentrated mass, gravity, and density, the singularity forms a huge depression, or cavity in space-time. Just as a man weighing 500 pounds would sink deeper into the mud than a woman weighing 100 pounds, collapsed stars also sink into the fabric of space time; and those with the most mass sinking so deep they form a huge hole, at the center of which sits that star's remains, which could be the size of a marble but with the gravity and concentrated mass of a billion suns.

As based on Einstein's general theory of relativity, once a star has collapsed it will create an intense supermassive gravitational field. In consequence, according to conventional wisdom, anything which falls into this gravity-laden cavity, including nearby stars and even light, can not escape such that surrounding this hole is blackness; i.e. a black hole.

It is believed that there are tens of millions of "stellar mass" black holes lurking in the inner and outer galactic arms and on the outskirts of the Milky Way galaxy, the gravitational remnants of dead stars, each with a mass anywhere from 10 to 25 more massive than the sun (McClintock, 2004; Schödel, et al., 2006). Then there are the supermassive black holes which have the mass of a million billion suns, one of which appears to sit at the axial center of this galaxy (Melia 2007). Then there are yet others which may have the concentrated mass and gravity of entire galaxies and perhaps others with the concentrated mass and gravity of an millions of galaxies (Joseph 2010).

For example, VIRGOHI21 (Minchin, et al., 2005) has swallowed all the stars of its galaxy and has the gravity of a small galaxy; an estimated total mass of about 1/10th the Milky Way; ten times more dark matter than ordinary matter; and is surrounded by vast clouds of hydrogen. Because of its galaxy-in-mass gravity, VIRGOHI21 has pulled up to 2000 galaxies toward it, creating the Virgo Cluster (Fouqué, et al., 2001). Thus, thousands of galaxies have been caught up in the vortex of this galaxy-in-mass gravity hole and now cluster about it.

The billion-light-years across "Eridanus black hole" may be typical of black holes which have the gravity-mass of millions of entire galaxies (Joseph 2010c). The Eridanus black hole sits like a giant black spider in an ocean of nothingness, having swallowed up all surrounding galaxies, gas, and light, including radiation from the Cosmic Microwave Background. Based on an analysis of the NRAO VLA Sky Survey data, Rudnick et al. (2007) in fact discovered that there was a significant and rather remarkable absence of galaxies even in the distant space surrounding this hole, in the constellation of Eridanus. Thus, the billion-light-years across "Eridanus black hole" must have consumed the gravity-mass of millions of entire galaxies all of which have been collapsed and concentrated into the singularity of this super-galactic hole.

Supermassive black holes not only suck up light, but may serve as mirrors into the past or act as windows into mirror galaxies. Smaller holes lurking in the arms of the galaxy may serve as smaller mirrors, mirroring nearby stars and their planets. Space-time may be littered with trillions of billions of mirrors into the past.

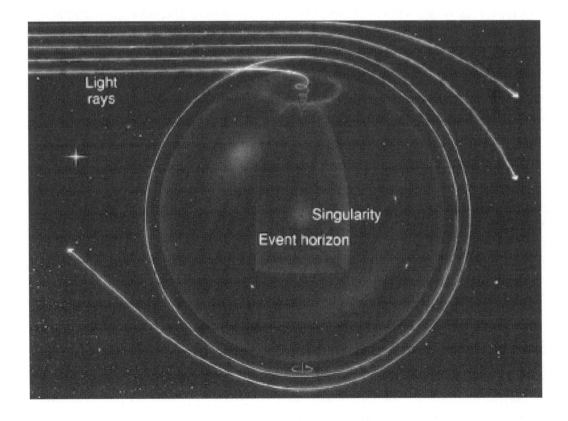

Quantum Physics of Time Travel

To journey into the past, the time traveler could dive his time machine into the hole, at which point he will be accelerated to faster than light speed and then enter the mirror at the bottom of the hole which leads to the long ago. Such a journey, of course, is not without its perils. The time traveler may be crushed to death, disintegrate into particles, or be irradiated into ash (Almheiri et al., 2013), with any remnants spewed out as disembodied particles (Hawking 2014).

The existence of black holes and their formation was predicted in 1939, by Robert Oppenheimer who argued that stars above approximately three solar masses would collapse into black holes and create singularities in the fabric of space-time. Adjacent to and surrounding the singularity would be empty space and enclosing that empty space would be a spherical event horizon; though some scientists, such as Hawking (2014) have proposed that "gravitational collapse produces apparent horizons but no event horizons."

Event horizons very depending on the mass of the singularity which punctured the hole into space-time. For example, the apparent black hole at the center of galaxy M87, has an estimated event horizon with a circumference of 56 billion kilometers (52 light hours).

Stars, objects, particles and astronauts just outside the event horizon are drawn toward it and their ability to escape from it become exceedingly difficult as they approach due to the incredible amount of gravity and the vortex that sucks in all surrounding space-time including light (Melia 2007, McClintock, 2004). As they approach the event horizon they will accelerate toward light speed giving off a red-shift light pattern indicating their incredible velocity.

According to conventional wisdom, whatever is captured at the event horizon will spin and circle round and round the circumference of the horizon at light speed; and as such, time will stop (Dieter 2012). Particles and objects that pass beyond the event horizon into the hole at first continue to circle at the speed of light, but as they are drawn down into the hole the forces of gravity accelerate their movement until they exceed light speed, and as such they are sucked into the past.

Broadly considered, there are several major types of supermassive black holes, including those with an electric charge, spin, and angular momentum, and those without (referred to as Schwarzschild black holes, named after karl Schwarzschild). General relativity predicts that any rotating mass will "drag" and pull space-time which also begins to circle around it (Einstein 1915a,b, 1961). Therefore, black holes with spin and momentum exert an organizational and gravitational effect on matter, just like a vortex in the ocean will drag surrounding water into the vortex. Almost every spiral galaxy is believed to have a supermassive black hole at its center (Bethe et al. 2003; Melia 2007, McClintock, 2004).

Supermassive black holes attract and organize stars which then circle and orbit around them, much in the same manner that water is drawn toward and then circles 'round a drain before disappearing inside.

Relativity, Space Time...

Just as the sun creates a depression and pocket in space time, thereby curving and dragging space-time into the hole it occupies, which in turn causes the planets to move along the curvatures, supermassive black holes at the center of galaxies have the same effect on all the stars of those galaxies. Stars littered throughout the spiraling arms of the galaxy, are like roulette balls, and orbit along the grooves in the fabric of space-time created by the curvature of space. However, the supermassive gravity of these super-massive black holes is so powerful, that not just space-time but entire stars are dragged into the hole.

Stars closest to the black hole have a greater orbital velocity compared to those further away (Ghez et al., 2005; Petrovskaya, 1994; Teerikorpi, 1989). Their velocity will increase as they come even closer to the hole. The light of the star will also begin to dim as light is suck toward the hole, and become red-shifted.

The point of no return the "event horizon." The "event horizon" is like the lips of a mouth, and completely encircles the outer rim of the black hole. Tidal forces are believed to be smaller at the event horizon and to decrease as the size of the hole increases, such that the bigger the hole, the more likely someone can survive.

If a space-time machine were to approach the event horizon its velocity would begin approaching the speed of light and it would be flung into the future. Upon reaching the event horizon, the space-time machine would have a velocity at light speed and time would stop. Those inside the time machine would experience an infinite "now." At some point the time machine, if still intact, would be sucked inside the hole and its velocity would increase well beyond the speed of light. And the same fate befalls all stars which are caught in the gravitational grip of a black hole.

From the perspective of an outside observer, the light associated with a star (or time machine) approaching and then falling into a black hole, would become dimmer and red shifted as it accelerates toward the event horizon until reaching an infinite red shift at the horizon. If the observer did not know there was a black hole, it would appear as if that distant star (based on the dimness of light) was accelerating (based on its red shift) and given its dimness (due to the black hole's capture of light) there would be an illusion that it is rapidly increasing its distance, becoming further and further away as it speeds up. However, although it is accelerating, it is not speeding further away but speeding into the future, due to the contraction of space-time; and the light associated with the star simultaneously becomes redder and dimmer. This is known as "gravitational red shift." Finally, the light would disappear, sucked down into the hole along with the star (and the space-time machine).

Vacuums and Negative Energy

There is considerable disagreement about the nature of black holes and what would become of a star or a time machine that happened to fall inside (Almheiri et al., 2013; Hawking 2014); with some scientists arguing that black holes do not even exist. For example, in addition to being ripped to pieces by tidal forces the Time Traveler may also experience what has been referred to as a Unruh thermal radiation; energetic photons which derive their positive energy from the vacuum within the hole, which gives the hole an energy density lower than zero; i.e. negative energy. These conditions create what is referred to as a Rindler vacuum which has negative energy density and negative pressure (Rindler 2001).

An example of a negative density quantum vacuum is illustrated by the Casimir force, a vacuum which becomes populated by virtual photons and negative energy (Casimir 1948; Jaffe 2005; Lambrecht 2002). Likewise, be it a super massive hole or a worm hole, the vacuum of the hole may create its own quantum vacuum which generates its own negative energy (Everett & Roman 2012). And negative energy has negative mass. This negative energy would be repellant wheres the positive energy would act as propulsion, thereby propelling the time traveler to beyond light speed; albeit, with negative energy and negative mass; a consequence of positive energy and mass being drained off to counterbalance the negative energy of the hole, thereby creating equilibrium as well as propulsion.

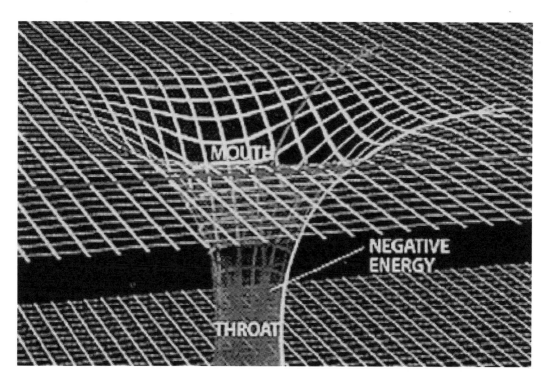

Specifically, if the time traveler were to enter the hole (and depending on the negative density of the hole or surrounding vacuum) she may lose mass and energy. become miniaturized due to length contraction, and emerge at the other end of the hole with negative mass and negative energy and then either grow larger in size, remain the same, or appear ghostlike, disintegrate or even self-annihilate if there is a collision with her positively charged double which is heading for the future while she is heading from the future for the past.

Even if there is no danger of inner hole negative energy density or positive to negative energy conversion, the time traveler and her time machine may still be stripped of electromagnetic energy which will radiate out into space (Hawking, 2005, 2014). However, this electromagnetic force (i.e. the basic unit of which is the quantum, the photon) being stripped of its particle, would have no mass, consisting only of a wave, thereby violating the particle-wave duality believed to characterize all matter; and leaving only a wave function as predicted by the Copenhagen interoperation of quantum physics (Bohr 1963; Heisenberg 1927). That is, upon exiting the hole and voyaging into the past, the time traveler may consist of negative mass (no mass) and a wave function.

Black holes spew particles and radiation (Giddings, 1995; Hawking, 2005; Preskill 1994; Russell & Fender, 2010). This raises the possibility that what is radiating out of one end of the hole, are the remnants of whatever entered at the opposite end of the hole. If, as Einstein and Rosen (1935) predicted, a mirror universe is at the bottom of the hole, stars, objects, time machines, etc. may enter from "the bottom" and emerge at the top as radiation. Black holes may be two way streets leading to the future and the past.

10: Microscopic Time Travel At the Speed of Light

Shrinking In Time

Myriad deadly hazards confront the time traveler, including premature aging, radiation, g-forces, and tidal forces if she were to tunnel through a black hole. Some scientists believe that everything that enters the hole will be compressed to the size of an atom and may become part of the singularity of the original collapsed star and everything else which has subsequently fallen into the hole (Bethe et al. 2003; McClintock 2004). If correct, it is unlikely a time traveler would survive.

On the other hand, since acceleration toward light speeds causes not just space-time but the time machine to contract, with everything remaining relative, it is theoretically possible that even when shrunk to the size of an atom that the time traveler and her time-space machine would survive a journey into a black hole. In fact, as predicted by Einstein's field equations and due to acceleration induced length contraction, the time machine and the time traveler and all immediately surrounding space-time may shrink to a nearly infinitely small size, less than a Planck Length ($1.61619926 \times 10^{-33}$ cm), which is the smallest unit of measurement. If this assumption is correct, then upon accelerating to light speed and before falling into the hole, so much energy would be compacted into that infinitely small amount of space that time-space would implode, punching a hole in space-time (Joseph 2010b).

In other words, even before falling into a supermassive black hole, the time machine may create its own miniature hole in space-time, as predicted by quantum mechanics as well as Newtonian and Einsteinian physics. The question then becomes, upon exiting the opposite end of the hole, would the time traveler, and space-time, return to a "normal size" relative to the mirror universe which exists on the other side? And, if time truly runs in reverse in the mirror universe, would the time traveler grow younger, would the aging process be reversed? And would she encounter herself as she journeys into the long ago?

Length Contraction

When an object is in motion and it accelerates, its mass increases as it absorbs energy, and it shrinks in the direction of its motion. Contraction in the direction of motion has been referred to as Lorentz contraction (Lorentz 1892, 1995; Einstein et al. 1923) and is predicted by Einstein theory of relativity (Einstein 1961). Everything shrinks and contracts in the direction of motion, as velocity increases. Although this shrinkage would be obvious to an external observer with a separate frame of reference, anyone inside the time-space machine would not notice any change. Everything inside the time machine would shrink to the same degree, including rulers, clocks, and any measuring device. At near light speed, the time traveler's length would contract to the size of an atom.

Length contraction is observable only by outside observers and is dependent on how fast an object is traveling with respect to those observers (Einstein et al. 1923). Many believe the height and width of an object in motion would not be effected, only its length which would be compressed in the direction the object is traveling. Therefore, the time traveler may become paper thin but retain her height and width. However, since increases in velocity compress space-time and increase mass, height would also be effected and reduced due to gravitational compression. Likewise, the shrinkage of time-space within atoms and spaces between and within molecules, and so on, would also reduce width as well as height and length.

The amount of length contraction can be calculated and determined by the Lorentz Transforms (Einstein 1961). For example, a 100 foot long time-space ship traveling at 60% the speed of light would contract by 20% and would become 80 feet in length. Presumably, its diameter would remain the same, though the likelihood is that all surrounding space including the diameter of the time machine would contract. If the time-space ship accelerates to 0.87 light speed, it will contract by 50%.

"Length contraction" can be expressed mathematically by the following formula: $E = mc^2/\sqrt{(1-v^2/c^2)}$, which is similar to the equation for time dilation (if one replaces the value of v for 0). As the value of v (velocity) increases, so does an object's mass which requires more energy to continue at the same velocity or to accelerate. Since energy can become mass, mass increases even as the object shrinks and contracts, thereby increasing its gravity which exerts local effects on the curvature of space-time. Not just the time machine, but space-time in front and surrounding the time machine also contracts.

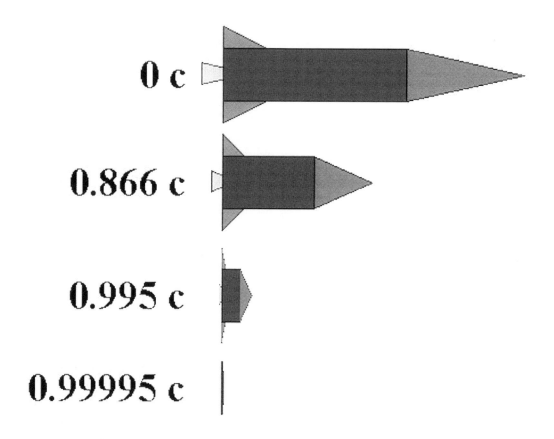

Humans as well as other objects and animals are also comprised of space-time. Because the time-space machine and everything inside and their atomic composition, are moving in the same direction, all contract to the same degree.

A time-space machine and the time traveler inside, are also comprised of particles and atoms. Therefore, the particles, atoms, and the space between atoms also shrinks as velocity increases. Hence, the spaces between biological (and non-biological) molecules and organs also shrinks and the curvatures and contraction of space-time affects not just external space-time but the time machine and the internal organs of the time traveler. Therefore the height and width of the time-traveler may also shrink and contract, since all the space between the molecules and the atoms of the body will contract effecting and shrinking the body (and the time machine), with the greatest degree of contraction taking place in the direction of motion.

According to Einstein (1961), an observer inside the moving object or traveling alongside at the same speed would not notice this contraction. It is only apparent to an outside observer with a separate frame of reference.

Relativity, Space Time...

Photons and the Paradox of Infinite Mass at Light Speed

Shrinkage and contraction exponentially increases as one nears light speed. According to some calculations, any object, the time traveler included, would shrink and contract to nothingness at light speed. And yet, as velocity increases, energy is added to the object which increases its mass. According to Einstein's theory of special relativity, any object which travels at the speed of light would shrink in size to nothingness but would also acquire infinite mass and would require infinite energy. However, this prediction may be based on the limitations of Einstein's theory which is not a theory of everything and breaks down when applied to events taking place at the atomic level and in spaces smaller than a Planck length (Bohr 1934; Heisenberg 1927, 1958). For example: how can an object moving at light speed have infinite mass and momentum and also shrink to nothing while maintaining some semblance of height and width since contraction is supposed only to effect length?

Photons which travel at light speed do not have infinite mass. In fact it is claimed that photons have no mass at all, for if they did, they could not travel at light speed.

A supermassive black hole does not have infinite mass (Bethe et al. 2003; McClintock, 2004, Melia, 2007) even though its event horizon is spinning at light speed (Brill 2012). Objects caught in the vortex of an event horizon do not acquire infinite mass, for if they did, then the hole and its horizon would acquire infinite mass. Likewise, objects which fall in the hole do not have infinite mass.

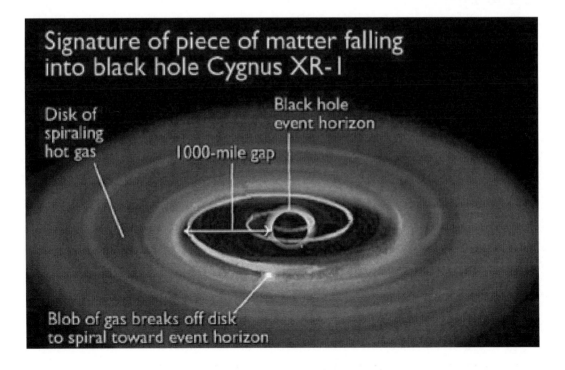

Quantum Physics of Time Travel

An object which is caught up in the vortex of surrounding space and the event horizon, and then swallowed by a black hole, may lose mass in the form of radiation which is expelled from the hole into the surrounding space medium (Giddings, 1995; Hawking, 2005; Preskill 1994; Russell & Fender, 2010). Therefore, it is possible that the time traveler and his time machine could lose all or most of their mass if they fall into a black hole. They could not, however, acquire infinite mass. As will be explained, the time machine and the time traveler may instead acquire negative mass and negative energy upon reaching superluminal speeds.

The solution to this theoretical conundrum appears to rest upon which theory one prefers, and of these theories, which is most applicable when describing microscopic phenomenon.

It is accepted by most scientists that relativity breaks down and quantum theory takes over when making measurements and observations of the microscopic world. Quantum mechanics describes effects at the scale of single particles, atoms and molecules whereas classical physics and Einstein's theory of gravity are more applicable when applied to large masses. Quantum physics takes over from relativity and Newtonian physics at the atomic and subatomic level.

Einstein's theories predict length contraction to sizes smaller than a Planck length. As based on quantum mechanics a particle shrinking to less than a Planck length gains incredible mass and energy, but not infinite mass (Joseph 2010b); and in consequence it can blow a hole through time.

According to Heisenberg (1956), "when quantum theory is combined with relativity, it predicts time reversal" in spaces smaller than 10^{-13}cm; smaller than the radii of an atomic nucleus--a Planck length. "The phenomenon of time reversal...belongs to these smallest regions."

However, in space smaller than a Planck length, and as these Planck length holes are created, mass has been stripped away the object which released the energy which creates the hole. As such, whatever enters these tiny spaces may emerge with negative mass and negative energy.

Therefore, based on relativity and quantum physics, it can be predicted that the time traveler, once she reaches light speed, may shrink to the size of a photon or smaller than than a Planck length. However, she does not gain infinite mass.

Unfortunately, all branches of physics break down when attempting to describe what takes place in spaces this small, though all agree the outcome may be the creation of a very small hole in space-time and the quantum continuum.

Planck Length and Microscopic Holes in Space-Time

In quantum physics, the smallest unit of space has a Planck length which is defined as 10^{-33} cm (Eisberg & Resnick 1985), about about 10^{-20} times smaller than the radius of a proton. Space smaller than a Planck length cannot be conceptualized by quantum mechanics or classical physics: Geometry ceases to exist, Cartesian coordinates, x, y and z, cannot be applied, and time ceases to have meaning (Garay 1995). Instead, a defining feature of these tiny spaces is gravity so powerful that it punches a hole in space-time (Joseph 2010b).

As predicted by Einstein's (1915a,b, 1961) theory of general relativity, any mass m has a length called the Schwarzschild radius, Sr. Compressing an object of mass m, to a size smaller than this radius Sr, generates tremendous gravity and immediately results in the formation of an almost infinitely small black hole in the fabric of space-time. Subatomic holes may be continually forming and then disappearing within spaces smaller than a Planck length (Joseph 2010b).

Two lengths of the Schwarzschild radius, Sr, become equal at the Planck length. Contained within a Planck volume is a Planck mass. Planck mass is also associated with quanta, which also refers to particles of light. Thus, in many respects Planck mass represents a division line between quantum and classical mechanics. In terms of time travel, Planck mass only becomes meaningful when applied to objects smaller than a Planck Length; objects which may then experience time reversal (Heisenberg 1958),

A Planck mass has a mass of about 22 micrograms (mP = 2.18×10^8 kg). Planck mass has sometimes been described as having gravitational potential energy which is generated between two masses which are separated by the angular wavelength of a photon. The gravitational potential energy can be derived mathematically when the Compton wavelength and Schwarzschild radius are equal. The Schwarzschild radius is the radius in which a mass, if confined, would become a black hole.

If we try to combine relativity with quantum physics, neither of which can accurately describe events taking place in space smaller than a Planck Length, and accepting that an object cannot be compressed to nothingness and retain height, width and mass, then it may be safe to say that a time machine upon reaching light speed, may contract to a size smaller than a Planck Length (Joseph 2010b) and then experience time reversal (Heisenberg 1958). Instead, of infinite mass, the now smaller than a Planck length time machine would have incredible mass, energy and gravity as suggested by relativity, Newtonian physics, and quantum mechanics. It would also have incredible energy; enough energy to blow a hole through time-space, thereby using up all its mass/energy, resulting in negative mass and negative energy.

Einstein's and Newton's theories of gravity both predict that if mass is shrunk to a subatomic space, its gravity will become increasingly powerful. Quantum physics tells us that in spaces the size of the Planck length, coupled with the corresponding Planck energy (10^{19} GeV), that the gravitational forces becomes so incredibly powerful (Eisberg & Resnick 1985; Smolin, 2002) that holes are created in space time. Hence, a space-time machine or any object with a Planck mass and whose radius is less than the Planck length, would have so much gravity that it could collapse surrounding space-time and create a black hole about the size of a Planck length.

Space-time, within the Planck scale, is subject to extreme uncontrollable quantum fluctuations, as it is continually being bent, folded, crumpled, and torn apart by these powerful gravitational forces (Bruno, et al., 2001). Holes in space-time and the quantum continuum are continually forming and disappearing (Joseph 2010b) and it is these holes will lead to a mirror universe where time runs in reverse.

As predicted by General Relativity, at least one hole may exist for every Planck length throughout space time. Therefore, all of space-time may be permeated by Planck length black holes which continually pop in and out of existence (Joseph 2010b).

Does this mean that particles and energy are continually leaking through these holes? The principles of quantum computing are based on this belief. Einstein (1939), however, argued that because of the principles governing the speed of light, particles could never enter these holes. And yet, Einstein's theories also predict that objects with sufficient mass and which are sufficiently small, will create these holes, whereas his theory of general relativity allows for superluminal speeds.

Therefore, based on general relativity and quantum mechanics, it can be predicted that at light speed, the time machine and the time traveler do not acquire infinite mass and shrink to nothing. Instead, upon approaching near light speed, the time machine and the time traveler may shrink to a size smaller than a Planck length, and they will contain so much concentrated mass, gravity and energy that they blow a hole in space-time. In other words, the time traveler and the time machine upon reaching light speed may create its own hole in space-time, and then emerge on the other side.

Negative Mass and Negative Energy and the Duality of Coming and Going

The time traveler must journey to the future, to reach the past. As the time machine accelerates toward light speed, space-time shrinks and curls up and contracts. The distance between the "present" and the future decreases and are pulled closer together. The time-space machine and the time traveler also shrink and contract and are propelled into the future which arrives more quickly compared to those back on Earth.

Upon reaching near light speed the time traveler shrinks to a size smaller than a Planck length, to a size smaller than a proton. Because the time traveler and the time machine have shrunk but gained mass and gravity, and due to the quantum composition of space-time in spaces smaller than a Planck length, they will blow a hole in space time and be expelled to the other side, and presumably into the past and the mirror universe at the other end of the hole as predicted by

Relativity, Space Time...

Einstein and Rosen (1935). However, upon blowing a hole in space-time, and upon tunneling through that hole, and as time-space reverses and continues in a negative direction, the time traveler may have negative mass and negative energy and remain microscopic in size.

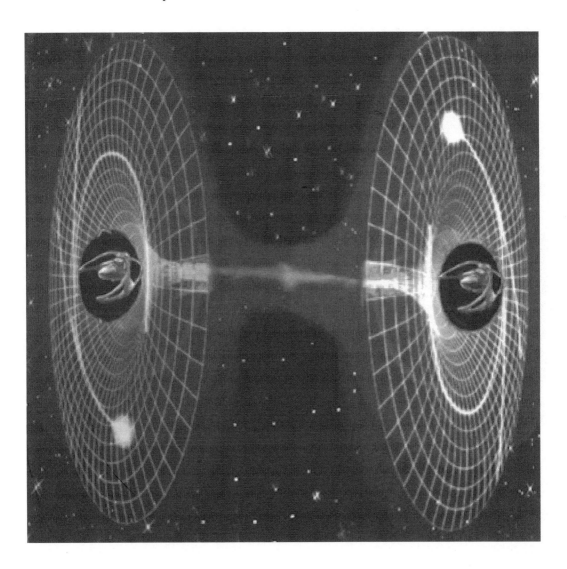

11: "Worm Holes" In Extreme Curvatures of Space Time

In 1935, Albert Einstein and Nathan Rosen proposed the existence of holes in the fabric of space-time which link this universe with a mirror universe on the other side. Presumably, in all respects, the mirror universe is exactly that: a mirror where everything is in reverse, including time. Since first proposed by Einstein and Rosen these holes have come to be known by different names each with their own properties, ranging from supermassive black holes at the center of galaxies to "worm holes" which link and tunnel through to distant regions of the cosmos thereby making faster than light space-time travel possible, at least in theory (Al-Khalili 2011; Hawking 1988; Thorne 1994).

If a space-time traveler was seeking a short cut between planets, stars, or galaxies, a "worm hole" between layers of space may allow her to arrive at these distant locations before any light which was emitted at the time she left. This is because light will bend and curve and follow the up and down crests and waves of the curvatures of space-time. Light has to travel a longer distance, whereas a worm hole offers a direct route. Instead of going over the mountain, the space-time traveler uses a tunnel which cuts through the mountain. Therefore, if she left Earth for the Andromeda Galaxy, and journeyed through a worm hole, even well below light speed, she would arrive well into the future, relative to time back on Earth.

However, if instead, a mirror universe, or mirror galaxy, or mirror planet lay upon the other side, then the space-time traveler may find herself in a universe where time runs backward. She would be tunneling from the present to the past. Conversely, if the time traveler was tunneling from the mirror universe, to this universe, which is also a mirror of what lays upon the other side, she may instead be propelled into the future.

Time is not an absolute. There are innumerable futures and pasts which exist simultaneously and which overlap, relative to different observers. For example, reflected sunlight that just left from Earth is now in Earth's past, but will not arrive until the future on Andromeda and the same principles apply to light from Andromeda on its way to Earth. Therefore, a space-time traveler journeying through a "worm hole" from Earth to Andromeda may arrive from the future relative to Andromeda, even though the Earth she left behind is in her past (Al-Khalili 2011; Hawking 1988; horne 1994).

Relativity, Space Time...

Gravity and the Layered Folding of Space

Time is a dimension, not in Euclidian space, but in "Minkowski space" (Minkowski 1909). Euclidian space consists of 4 spatial dimensions which include movement and geometric space; but none of which encompass time. By contrast, in "Minkowski space" which is incorporated within Einstein's special relativity, time is the 4th dimension (Einstein 1961). More specifically, 3 of the Euclidian dimensions of space are combined with a dimension of time thereby creating a four-dimensional manifold known as "space-time."

Einstein's theories predict that the entire universe is curved, whereas his field equations predict that anyone traveling in a wide enough circle across the cosmos will return to their starting point, such that the present leads to the future which leads to the past (Gödel 1949a,b). Because of gravity, time-space can also be bent backwards in a circle, as happens with whirlpools and eddies along river banks where water flows in a circular motion. If the implications of Einstein's theories are correct, then the river of time is bent round in a circle and it has no beginning or end.

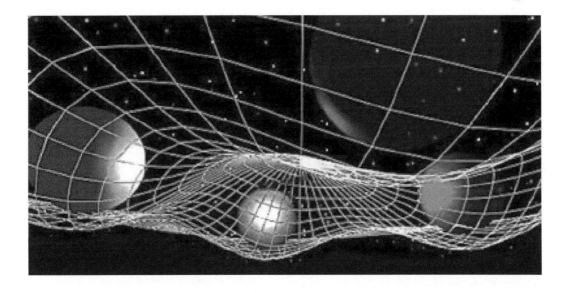

The curvature of space-time, time, and light, is a consequence of gravity which can can warp, shrink, compress, and punch holes in space-time (Al-Khalili 2011; Hawking 1988; Parker & Toms 2009; Ohanian & Ruffini 2013). The gravity of the sun creates depressions and curvatures in time-space which are more severe than those of Jupiter whose own curving vortex is stronger than that of Earth; and the gravity of these planets, the sun, and entire galaxies can bend and curve light, creating even multiple images which appear side by side; and the same may be true of space-time which can be split apart.

Gravity is relative (Einstein 1915a,b, 1961). The mass and gravity of this solar system differs from the 500 million other solar systems in this galaxy. For example, the mass of the Milky Way Galaxy (5.8×10^{11} solar masses M) is less than the Andromeda Galaxy (7.1×10^{11} solar masses M) with its one trillions stars (Young 2006). In consequences, time flows differently in these galaxies which also create differential pockets within space-time.

Throughout space-time, gravitational fields vary and differ in strength (Parker & Toms 2009; Ohanian & Ruffini 2013). Although space-time is curved, the curvature is lumpy, with wave-like crests and valleys in a turbulent sea of quantum foam such that distant regions may be drawn nearly side by side, like the flotsam rising up and riding adjacent waves in a turbulent sea. In some regions of space-time the effects of gravity are so powerful that distant regions fold and curl up side by side. Just as distant objects on the surface of a calm sea can be temporarily brought close together by wind acceleration or tidal forces, riding high on two adjacent cresting waves, the distance between stars and galaxies can be dragged closer together when gravitational forces cause tremendous crests and valleys in space time--a consequence of surrounding space-time being dragged down into the cavities these stars and galaxies form with their mass and gravity.

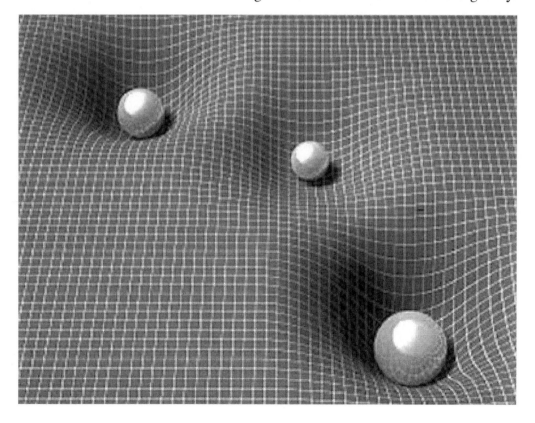

If a rubber sheet were stretched from one end of the Milky Way Galaxy to the other, then there would be billions of gravity pits representing all its stars, planets, and moons, with these each of these cavities varying in size, all of which individually and collectively effect the geometry of space-time and the trajectory of light (Carroll 2004). Each of these gravity holes curve space-time to varying degrees, such that space-time geometry contracts and is displaced with some areas of space subject to tremendous amounts of torque, pressure and tensions, and others less so.

For example, if there is 12 foot of rubber sheeting, and the Milky Way galaxy and the Andromeda galaxy were placed 4 feet apart at the center they would create depressions and cavities and drag the surrounding rubber sheeting toward them and down into the cavities they create. Moreover, they would be pulled toward each other; just as two balls dropped from different heights are pulled toward the center of gravity as they fall. They would be pulled toward their combined center of gravity.

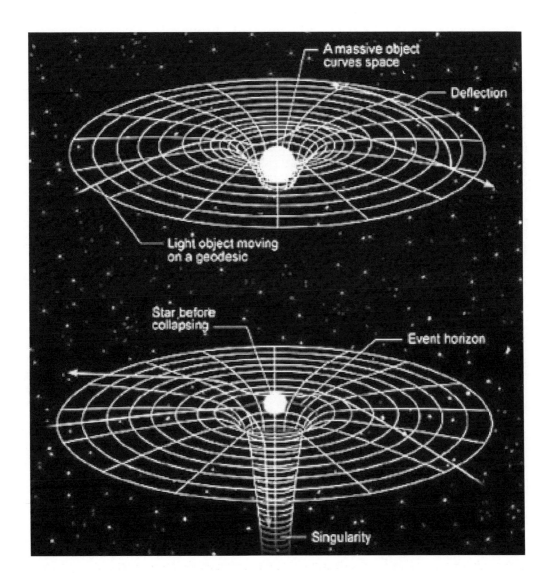

Quantum Physics of Time Travel

If Andromeda creates an indentation 2 ft deep and the gravity of the Milky Way a pocket 1 ft in depth in space-time, then the sheeting which had been on the surface and between them would be dragged into these pockets of space-time and the two galaxies (the caverns they created) would end up closer together, although the distance, when measured in the amount of rubber sheeting, would stay the same. This is because some of that sheeting (space-time) has been dragged down into the concave gravity spheres their mass and gravity created.

Despite this extreme curvatures and folding up of space-time, if a rocket were sent from the Milk Way to Andromeda, it would have to travel along the layer of space-time where the Milky Way is located, then up and over the layer leading to the pocket in space-time created by the Andromeda galaxy, then down along the layer of space-time into the crater formed in the fabric of space-time by Andromeda.

Moreover, because each has ballooned downward into the indentations they made, the pockets created by these two galaxies are separated by two layers of space-time which form the inner lining of the deep spherical pockets created by each galaxy. Although its the same layer of space time, it has been folded downward, side by side, creating two layers and a vacuum between them.

Holes in Curvatures of Space-Time

Light Follows Curvature of Space-Time

However, these great downward curvatures are not apparent to ground level observers due to the nature of light which also bends. Light travels up and down these crests and valleys in geodesics, which is the "straightest" path

Relativity, Space Time...

available. Even Earth, from a local, ground-level view, appears flat, only to suddenly disappear at the horizon. Just as a ship can't be seen after it crosses over the horizon, one can't see the curvature of Earth except from a height well above Earth. The same applies to the curvature of space and the depressions formed in space-time by the gravity of planets, stars, and entire galaxies--it would require an infinite "god's eye" view.

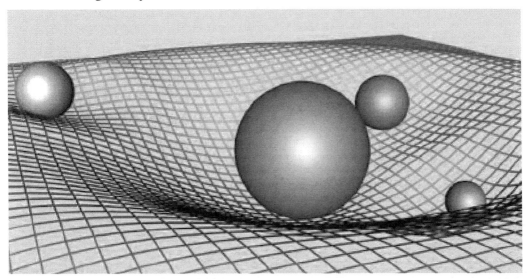

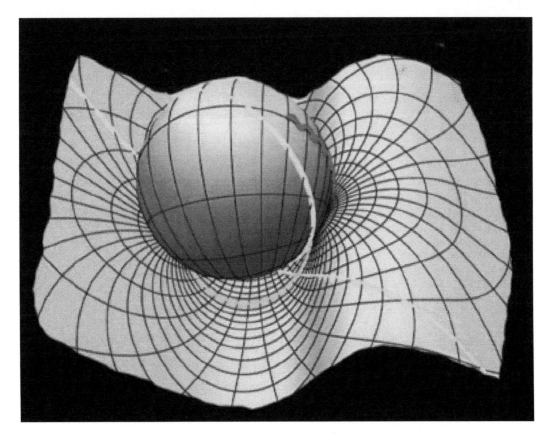

Quantum Physics of Time Travel

For example, a turbulent sea with waves cresting and crashing together with all the fury of the cosmos, would appear flat, albeit splotched with foam, when viewed from hundreds of miles above the planet. It is this same quantum foam which can bubble and ballon upward or downward into space-time, some of which end up side by side, like two soap bubbles, separated yet linked by tenuous membranes. Membranes, however, are porous, and gasses and other substances may pass between them through holes and perforations which link them.

"Worm Holes" In Space-Time

Due to gravity space-time not only curves, but it may dip or fold up, like two waves where the crests are drawn together at the top and the troughs/valleys are drawn together at the bottom, thereby shrinking the distance between the waves. And gravity creates depressions which may balloon together, separated by layers of space-time.

It is this ballooning and side-by-side layered folding which also makes faster than light space-time travel possible. These powerful and differential gravitational forces create pits and pockets such that distant locations are, relatively, side by side, separated only by the folds and curvatures in space-time which form the spherical cavities punched into the quantum continuum and time-space (Al-Khalili 2011; Parker & Toms 2009; Ohanian & Ruffini 2013; Thorne & Hawking 1995).

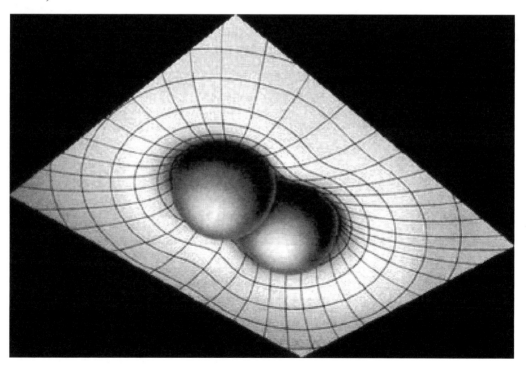

Relativity, Space Time...

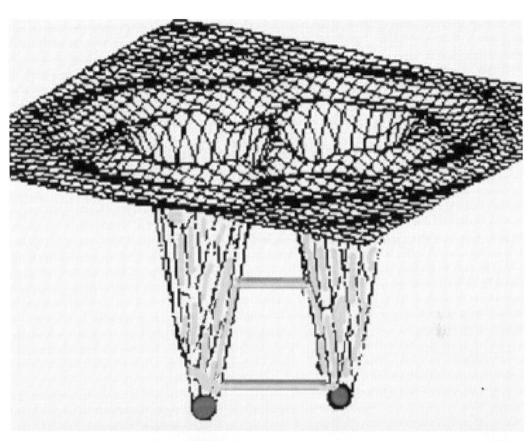

In consequence, it is possible to tunnel between these cavities and to arrive at the other side in less time than it takes the speed of light which must travel the more distant curving route.

For example, the distance from Manilla in the Philippines to Cuiabá in Brazil is 12,223 miles. However, the planet is curved and these two nations and cities are antipodal to each other, on opposite sides of the Earth. If one were to tunnel from Manilla through the center of the Earth to Brazil, the distance would be shortened by about about 35%, from 12,223 miles to 7,900 miles.

Likewise, since, according to Einstein, space is curved, if one could tunnel through space-time, the distance between two planets, two galaxies or two universes on opposite sides of each other, would significantly decrease. Hence, if a space-time traveler were to tunnel between the layers separating the ballooning gravity compartments created by galaxies, stars, and planets, then instead of following a curving straight line, she would be able to journey between galaxies or universes using a route that is significantly shorter. In fact, after the tunnel has been carved between these depressions in space-time, she may be able to journey from one galaxy to another faster than the speed of light which would be taking the longer curving geo-disc route up and down the valleys of time-space.

A tunneling hole would make it possible for the time-space traveler to pass directly from folded layer to the next folded layer, that is, from the lining of one pocket to the lining of the other, using the shortest route possible which would include the vacuum between the folds. For example, imagine a flat map which is folded or rolled up such that Argentina and China are face-to-face. If that map was the actual size of this planet, these two nations would be over 12,000 miles apart. However, if the map was folded or rolled up, and if one could simply tunnel or leap across the empty space between the folds, instead of following the contours of the map, the distance might only be a few hundred miles. The same principle can apply to extreme curvatures and folding of space.

Of course, leaping across or tunneling and digging down into space-time is not possible with current technology. Therefore, the time traveler must locate a rip, puncture, or perforation in the fabric of space-time which is so deep it creates a tunneling hole leading from the surface of one curving layer to the folded surface of the other portion of the layer which lines the cavities which are adjacent to and created by the gravity of various galaxies, stars, and even planets. There are dangers, of course, including emerging stripped of positive energy and consisting only of negative energy and negative mass .

Worm Holes and Negative Energy in Space-Time

No matter how mild or extreme the curvature or folding, the space in between folded up layers of two adjacent cavities created by the mass/gravity of various stars or galaxies will likely consist of a vacuum of space permeated by negative energy. Imagine two bowling balls placed near the center of a rubber sheet. Both balls will cause the rubber sheet to sag beneath them so that the same sheet is dragged into and lines the cavities formed by the two balls. Although its the same rubber sheet, the adjacent areas where the two cavities are side by side creates a folded up layer, a bilayer, and between the bilayer and the two cavities, is space, a vacuum.

Empty space (vacuums) is also warped by gravity waves and by moons, planets, stars, and galaxies; as predicted by the "Schwarzschild solutions" which describe the curvature of empty space adjacent to massive objects (Einstein 1961).

Therefore, vacuums and space-time are subject to curvature and extreme tidal forces, pressures and tensions which differ depending on gravity. Moreover, these folded up layers and the vacuum space between them, are also differentially effected by the mass and gravity of different planets, stars, and galaxies which created the depressions and pockets in space-time. One folded up layer may also be subject to greater tensions than the other due to greater gravity, creating electromagnetic and quantum inequalities in energy densities.

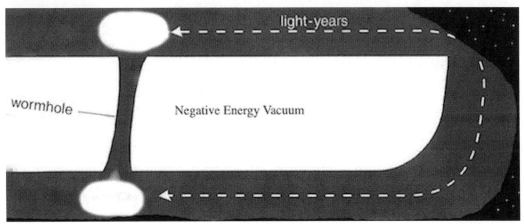

Quantum Physics of Time Travel

These varying dynamics and differential pressure gradients on the surface of and within and between these curled and folded up layers can differentially torque adjacent folded up layers and can create pockets of negative energy within the spaces (vacuums) between them. In consequence, holes may open up within these folded up layers of space-time, creating pores, or passageways, which serve to equalize these differential pressures and energy fluxes including and especially those created in the vacuum between these layers--like letting air out of a tire. Theoretically, the formation of these pores and tunneling holes could also allow a space-time traveler to pass through them, entering the hole on one side of a folded up layer passing through the space between them, and exiting the hole in the other, thereby quickly traveling to whatever planet, galaxy, or universe is on the opposite side (Al-Khalili 2011; Morris & Thorne 1988).

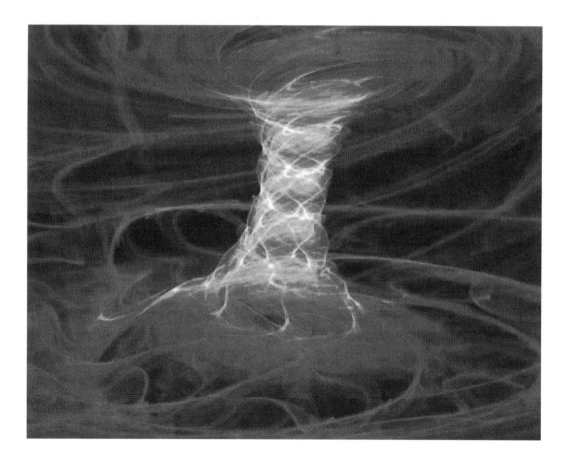

Again, this could be likened to tunneling from China to Argentina, with the crust of the Earth beneath China forming one layer, and the crust beneath Argentina the other. Although the crust of the Earth circles Earth and is the same crust, it nevertheless forms two layers as it circle rounds, and each layer must be traversed if one is to tunnel between them. Likewise, the space-time traveler would be ferrying through two folded up layers: the concave pocket of space-time created by the gravity of one galaxy, through a vacuum, and then through the concave layer formed by another galaxy. Although its the same space-time twisted in a circle, the opposite sides of that circle of space-time, when juxtaposed, form a bilayer; and because of the pressure that builds up in the vacuum between them, these bilayers may spring a leak, in the form of a "worm hole."

Time Machines Through the Tunnel of Time

The advantage of taking a journey through these tunnels of time, is, time. For example, the Andromeda galaxy is 2.5 million light years from Earth. Even at 99.99999 light speed, it would take over 2.5 million years to take a one way trip to Andromeda. However, because of the curvature of time and the proximity of these pocket-like bilayers, the gravity cavity surrounding Andromeda may be (hypothetically) only 10,000 or 1,000 light years distant from the depression made by the Milky Way.

Perhaps more practical in terms of time and distance, would be the Large Magellanic Cloud which is 160,000 light years distant, or Canis Major Dwarf Galaxy which is a mere 25,000 light years away but which may be only a few hundred if one were to take a direct route through these time tunnels. Or even better, Proxima Centauri, the nearest solar system to our own, and which is only 4.24 light years distant, but likely even closer if the pockets of gravity created by this solar system are right next to our own.

Upon entering these time tunnels which lead from one folded layer to the next, a journey to a distant solar system may take only a few days or weeks; if traveling at near light speed. Upon exiting these worm holes, the space-time traveler would not only arrive at her destination more quickly than light beams journeying from Earth, but far into the future of Earth. Moreover, upon arriving at Proxima Centauri, or Canis Major or Andromeda, the space-time traveler would be able to gaze backward in time at an Earth which existed before she left on her journey.

For example, if our time traveler took the direct route via worm hole and arrived on a planet orbiting Proxima Centauri, and the trip took only a few days, she could gaze into the skies of that alien world and look upon Earth from 4.24 light years ago. She would be in the future relative to those back on Earth. However, the past of Earth (the previous 4.24 light years) would not arrive on Proxima Centauri until the future. Thus, the "present" for the time traveler on Proxima Centauri would overlap with both the future and past of Earth.

However, in planning for a worm-hole journey to another world, there is not only the problem of finding a hole larger enough to allow entry (and knowing where it leads to), but a means of safely and quickly taking a journey through these tunnels of time. The hole may close up while the time traveler is still in it.

12: Worm Holes, Negative Energy, Casimir Force And The Einstein-Rosen Bridge

The possible existence of holes in the tissues of space-time is based on the work of Einstein and Nathan Rosen (1935), and initially these holes were referred to as the "Einstein-Rosen Bridge." According to Einstein and Rosen, at the top of the hole is a "mouth" and at its center is a "throat." The hole does not lead to a bottom but opens at the other end, forming another "mouth" and which leads to a mirror universe. However, these theories were eventually abandoned and the Einstein-Rosen bridge dismissed as a mathematical anomaly.

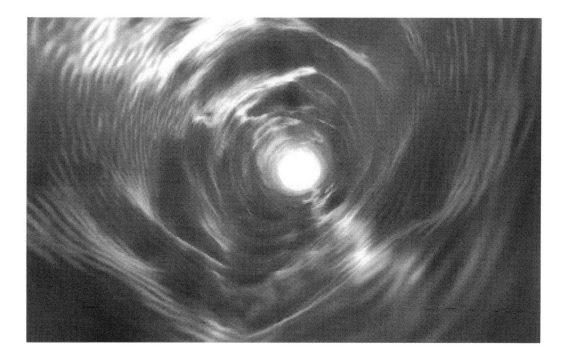

Rosen and Einstein's ideas and theorems were resurrected in the 1960s by Robert Fuller and John Wheeler (1962) who saw them as a mathematical requirement for proving the existence of supermassive black holes. They are also a key component to the Reissner-Nordstrom solution which describes an electrically charged black hole. John Wheeler in fact coined the term "black hole" to symbolize its two central characteristics: emptiness and blackness.

The Einstein-Rosen bridge and the holes they visualized are also sometimes referred to as "worm holes." "Worm holes" differ from supermassive "black holes" but also share many of the same features, such as negative energy densities (Everett & Roman 2012).

Wheeler and colleagues came to the conclusion that these "holes" would be unstable and would collapse so quickly that even a light beam would never make it through but would be pinched off at the throat (Fuller & Wheeler, 1962; Taylor & Wheeler, 2001). Wheeler and others have suggested that the curvatures of the walls inside these worm holes may smack up against one another, leaving no space between them.

"Worm holes" are not necessarily worm-size but can be planet-size or larger. It has also been proposed that worm holes (like "black holes") have an event horizon and any object approaching a worm hole would get caught at the horizon and could never make it inside. The "Einstein-Rosen Bridge" was thus a bridge to nowhere and no one could get across.

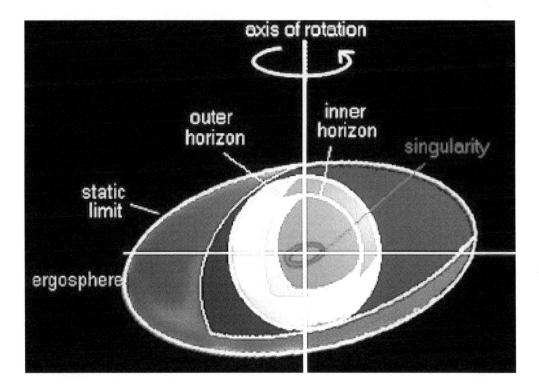

Others have proposed various worm hole and black hole geometries which could make it possible to tunnel from one galaxy or universe to another (Al-Khalili 2011; Hawking 1988; Thorne 1994). In 1963, Roy Kerr proposed a solution to Einstein's equations which predicted that a collapsing star would be rotating as it collapsed. Because of the rotation it would not collapse to a singularity, but become compressed into a ring which allows passage from one end of the hole to the other (Kerr 1963; O'Neill 2014). Therefore, anything passing through the throat would not be subject to infinite curvature or infinite gravity, but finite gravity. A time traveler falling into the hole would be shot right though to the other side. The Kerr black hole is a hole that acts as a gateway to the mirror universe. Unfortunately, Kerr holes are so unstable they might collapse before the time traveler could exit.

Mike Morris and Kip Thorne (1988) proposed that "traversable worm-holes" would have no event horizon, no infinite curvatures, no tidal forces, and could be easily traversed if propped open and lined with "exotic matter" which would have a repulsive gravitational effect--meaning, other mass would fly away from it. The inner walls of the worm hole would repulse one another and remain open. Their proposals, however, violate the so called "weak energy" hypothesis: neither matter or energy can ever be negative and can never have less than the mass or energy density of empty space. The weak energy "hypothesis" however, is just that and it has been proved wrong (Jaffe, 2005; Lambrecht, 2002).

Morris and Thorne (1988) proposed that their "exotic matter" should be in the throat of the worm-hole, forming, presumably, a symmetric ring of exotic matter which would repulse the walls of the hole, keeping the worm hole open and the walls far apart. Unfortunately, any space-time traveler would have to pass through this ring, and suffer the unknown effects of this exotic matter. As a "solution" Thorne and Morris suggested inserting a vacuum tube through the throat which could act as a shield. Thorn has also suggested towing these holes to different locations in space-time thereby making it easier to journey from one location to another in faster than light speed (Thorn & Hawking 1995). Unfortunately, Thorn has never explained what this "exotic matter" is, or how it would be possible to cart these holes from place to place.

The "worm hole" theories and proposals that have received the most attention from the scientific community and the media are in many respects a generalized modification of the standard "black hole" theory (Fuller & Wheeler 1962; Taylor & Wheeler 2000); i.e. the gravity of a mass compacted to a singularity punches a hole in space-time; or, a spinning vortex of a collapsing star drills the hole and never forms any singularity; or exotic matter creates and maintains the hole.

Nevertheless, "worm holes" should be distinguished from "black holes" and although rather fantastical scenarios have been published in various books and scientific journals about their nature and creation, only a few scientists have provided explanations as to how and why a "worm hole" would form between planets, solar systems or galaxies and how and why the hole would link them.

For example, John Wheeler proposed what he calls "spacetime foam" which can form a labyrinth of tunnels and holes the size of a Plank length filled with positive and negative (virtual) particles (Wheeler 2010). Virtual particles are continually popping into existence, permeating space with quantum activity such that there is no emptiness, but only fluctuations involving frenzied activity. An example of this is a quantum vacuum as illustrated by the Casimir force which becomes populated by virtual photons and negative energy (Casimir 1948; Jaffe 2005; Lambrecht 2002).

These quantum holes and the virtual (negative mass) particles within them live on borrowed time. Because they disappear so quickly, lasting no longer than Planck time, they have been referred to as "virtual" (Wheeler 2010). There would not be much time to travel through a tunneling hole in space-time foam.

How Holes Form Between Layers in Extreme Curvatures of Space-Time

Space-time is not a solid but an expression of the quantum continuum which consists of a frenzy of particles and waves that are perceived and experienced as substance, form, and the various dimensions of the universe (Bohr, 1934, 1958, 1963; Dirac 1966a,b). Because it is not a solid but consists of space, time, and vacuum, space-time may contract and can be stretched, twisted, ripped apart, and may be permeated with perforations and holes which tunnel to distant locations (Fuller & Wheeler, 1962; Wheeler 2010).

When a single layer of space-time folds up and over or curves around itself, thereby forming bilayers with an empty space (vacuum) in between, differences in concentration gradients in gravity and energy may create pockets of pressure and forces similar to those which drive osmosis and diffusion (Wilf et al. 2007). As to how long these holes remains open depends on the pressure, concentration, temperature, gravity, electrical charges and other factors involving not just the folded layers (which form a bilayer) but the vacuum in between the bilayers and differential energy gradients and negative versus positive energy fluxes within and between them.

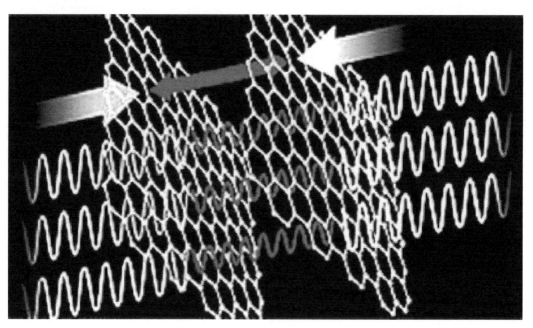

Figure: Casimir vacuum with exchange of negative and positive energy

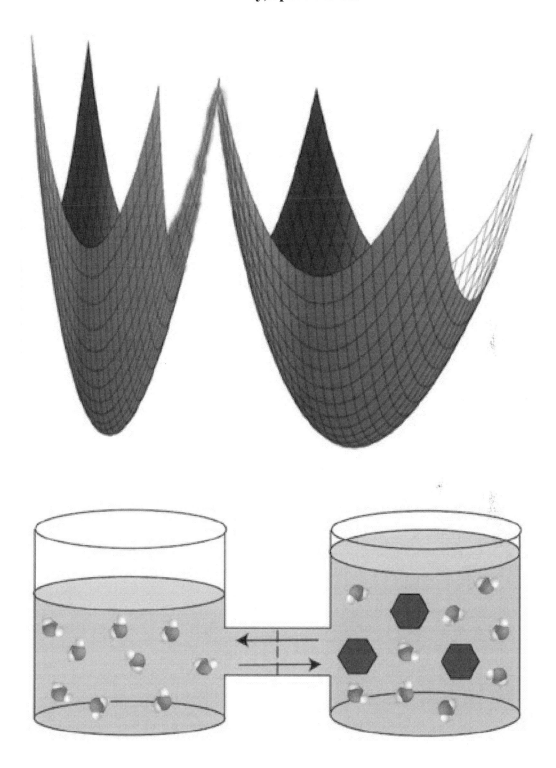

Figure: Osmosis and diffusion

Differential gravity and energy densities, and the consequence of dragging space-time into respective gravity-bubble cavities can create so much tension that the tissues of space time are weakened making specific areas subject to ripping, tearing, and hemorrhaging; like when too much air in a rotating tire heats up and expands and a hole bursts open allowing air to escape. When negative vs positive energy levels within the vacuum and the pressures above or beneath one layer or both become extreme, any weak points formed in space-time hemorrhage and holes pop open thereby allowing passage through these layers and the vacuum between them; similar in some ways, to osmosis (Wilf et al. 2007).

Consider, for example, Andromeda and the Milky Way Galaxies, each of which causes extreme curvatures as they sink deep down into space-time--just as two fat men standing in a pool of mud will sink into the wet ground. However, unlike the two men where some of the mud beneath their feet plops out onto the surface of the pool of mud, these two galaxies drag space-time down into the cavities they create. In consequence, the distance between them also shrinks as surrounding space is pulled down into the depressions they've made. These spherical balloon-like cavities may be nearly side by side, separated by lengths of the layer of space-time dragged down into each of these gravity pockets, with the layer forming the wall of one cavity separated from the wall of the other cavity by the empty space between them; like two adjacent pockets woven from the same cloth. One layer of space-time stretched down into and lining the two cavities which are relatively side by side, creates a bilayer and between the bilayer: a vacuum of empty space.

However, when a vacuum forms between layers, a negative pressure density develops and the vacuum comes to be permeated by negative energy (Casimir, 1948; Bressi et al., 2002; Jaffe 2005; Rodriguez et al., 2011). Exterior to the layers surrounding the vacuum, the opposite sides of space-time will be subject to a positive energy density, which may be greater on one side than the other. These differential pressure gradients can exert so much stress that holes may open up and tunnel between those layers thereby releasing pressures building up within the vacuum and between them.

Moreover, the layer of time-space lining one gravity-cavity may also be differentially effected by gravity and electromagnetic activity triggering changes in polarity and negative vs positive energy fluxes which propagate between layers, thereby effecting the vacuum between them. For example, one layer may have a greater positive energy density than the other, or the vacuum (empty space) between the bilayers may be permeated by pockets of positive and negative energy (Everett & Roman 2012). In consequence, holes, or channels, may pop open to release this pressure and to equalize energy gradients.

Differential gravity and electromagnetic activity creates an imbalance in electric fields which can lead to electromagnetic diffusion, conductance of electrical charges (Pollack & Stump 2001; Slater & Frank 2011) and the opening up

Relativity, Space Time...

of holes in the bilayers of space-time so as to equalize charges and polarity. Different positive energy densities on either side of a folded layer of space may trigger electromagnetic permeability and the conductance of charged particles from one side to the other to establish electromagnetic equilibrium (Pollack & Stump, 2001; Slater & Frank 2011). Holes may be forced open from the outside, as well as from within the vacuum due to the interactions of different electric fields and energy densities on either side of the folded space-time layer.

The Casimir Vacuum

The vacuum between a folded up layer of space-time will most likely have a negative energy density, or, perhaps pockets of negative and positive energy (Casimir, 1948; Bressi et al., 2002; Jaffe 2005). In a series of experiments first performed by Hendrik Casimir in 1948, it was found that the "vacuum" between two uncharged silver plates aligned together developed a negative energy density, i.e. less than zero. Thus energy must be added to reach zero. By contrast, in a vacuum that was not surrounded on two sides by plating there was no negative energy and the energy density was already zero.

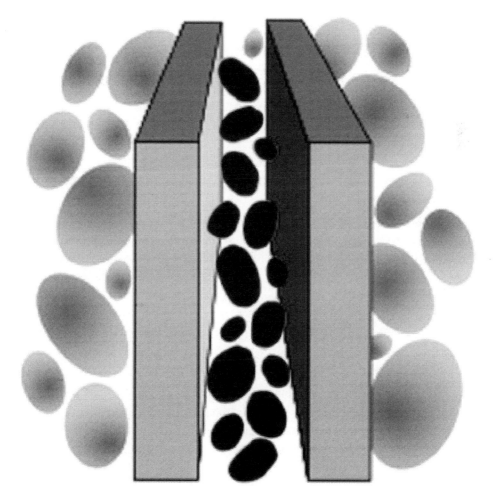

Casimir (1948) found that these two uncharged mental plates, when aligned close together with the vacuum between them, demonstrated an attractive force due to vacuum fluctuations; the plates were drawn together. Further, the negative energy density increased as the two plates were pulled closer together.

The same "Casimir Effect" may also characterize the vacuum between folded or curled up layers of space-time; i.e. there is a buildup of negative energy and the layers of space-time are drawn closer together.

In the Casimir "vacuum" there is positive pressure in two directions parallel to the plates, but a large negative pressure inside which sucks the two together--and the closer the plates (or layers) the greater the negative energy density between them (Bressi et al., 2002; Jaffe 2005; Rodriguez et al., 2011). Normally, a layer or vacuum with positive energy density would exert pressure outward (like air in a tire). A negative energy vacuum generates inward pressure or suction. Thus, in response to extreme curvatures of space-time, adjacent layers (be they like the crests or valleys of two waves, or bilayers formed when two depressions are drawn together) would also be pulled closer together by the suction of the negative energy vacuum. When adjacent layers are pulled closer together this puts more pressure on the vacuum between them which in turn increases the negative energy density within (Casimir, 1948; Bressi et al., 2002; Jaffe 2005; Lambrect 2002).

Space-time is not a solid. Suction from a vacuum between layers would suck positive energy outside the layers into the vacuum, thereby creating holes large enough to allow the passage of positively charged particles and mass; acting similar to a semi-permeable membrane. As negative energy density increases it would cause holes to pop open in the outer layers thereby allowing negative energy to escape and decreasing the vacuum state within; just as a garden hose with a kink sprouts several holes as the pressure builds up thereby allowing water to burst from the holes. Holes open up so as to establish equilibrium; again, similar to osmosis.

For example, if a container of water is divided by a semi-permeable membrane, and 6 ounces of salt were poured on one side and 1 ounce on the other, it would create an imbalance with osmotic pressure which is higher on one side and lower on the other. High osmotic pressure provides a high energy density and thus the energy to force salt across the membrane, from the side with the higher concentration to the side with lower, until the different concentrations would equalize thereby driving the osmotic pressure to zero (Wilf et al 2007). The same amount of salt eventually end up on both sides of the membrane.

These kinks in the curvature of space time would also give rise to vacuum fluctuations which differ in the squeezed areas vs those further apart (Everett & Roman 2012). Those areas squeezed and drawn closer together would contain pockets of increasing negative energy density whereas those further way may contain pockets of positive energy density. Positive energy is more powerful than negative energy, with positive being attracted and negative repulsive, thereby increasing pressure against the fabric of space-time. In consequence, one or both folded or curled up layers would hemorrhage thereby allowing negative energy to escape and positive energy to rush in through the holes between them.

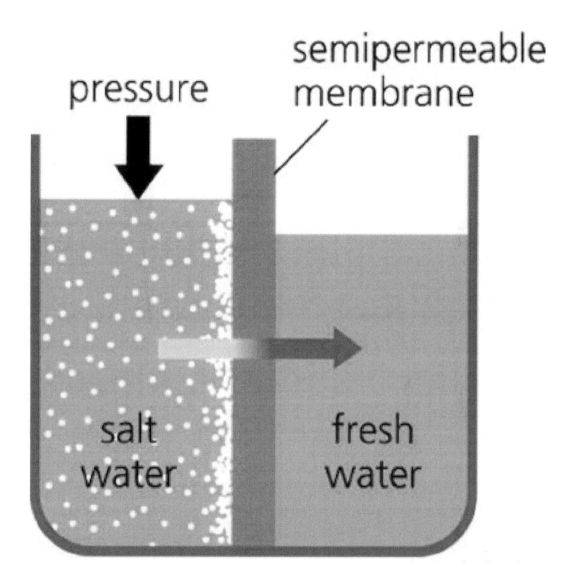

Likewise, in regions where space-time is folded or curled together, differences in concentration gradients in gravity and positive vs negative energy density or quantum inequalities in electromagnetic activity may puncture these layers and force holes to open. Differential gravity and electromagnetic activity also creates an imbalance which can lead to electromagnetic diffusion and the conductance of electrical charges (Pollack & Stump, 2001; Slater & Frank) and the opening up of holes in the bilayers of space time. Holes may be forced open from the outside, not from within the vacuum. Once the holes are formed, energy will stream through the hole and in and out of the vacuum between the bilayers, thereby allowing for the diffusion of positively charged molecules and larger objects which can pass from one side to the other and into and out of the vacuum.

The same principles can be applied to a time traveler. The time traveler and the time machine would consist of positive energy and positive mass. Therefore a time traveler could also be sucked through these holes and deposited on the other side--though there is a danger he may remain trapped inside the vacuum if the hole closes up or he may be stripped of positive energy and mass.

For example, as positive charged particles pass through the vacuum between the bilayers, they will lose positive energy, thereby equalizing the energy density outside and inside the hole. The hole then closes up. However, positively charged particles, in losing their positive charge as they pass through the hole may have a negative charge when they emerge on the other side (Everett & Roman 2012). If positive charges enter both ends of the hole, and exit the opposite ends as negative (after having their positive charge stripped away), this will have an equalizing effect, such that the charges inside the vacuum of the hole and those outside the hole are in equilibrium; at which point, the hole may grow smaller in size and then close up.

Therefore the tube-like vacuum between folded up layers of space-time may contain pockets of positive and negative energy (Everett & Roman 2012). As the "attraction" is repulsive, and as positive energy can have mass whereas negative energy would have only negative mass, the pockets of positive energy can push against the negative. As the negative energy is repulsive and has no where to go except out, these oppositional positive vs negative forces may cause the fabric of space-time to hemorrhage and leak negative energy which is pushed out by the pockets of positive energy inside the vacuum; or as a consequence of the two layers being drawn so close together than negative energy has no where to go except out.

Thus any the particles and radiation spewed out of a worm hole may be a consequence of the negative energy density within the vacuum between bilayers. These represent the remnants of positively charged particles and mass which originate from outside the layers of space time from the opposite end of the hole, or from negative energy from within the vacuum which runs the length between the bilayers of space-time. Once equilibrium is reached, these "worm holes" close up because they have equalized the negative and positive energy densities within the vacuum which forced these hole to open in the first place (Everett & Roman 2012). However, as the nature of these vacuums is to generate negative energy, the cycle of holes opening and then closing may infinitely repeat itself.

Gravity, Quantum Inequalities and Negative Energy

Gravity also produces negative energy (Everett & Roman 2012). For example, the gravity of Earth produces a field of negative energy which surrounds the planet; a consequence of dragging virtual photons toward the surface. Jupiter, the sun, entire galaxies and supermassive black holes in space-time produce and generate tremendous amounts of negative energy. And this negative energy may be forced into the depressions, pockets, spaces and bilayer vacuums that permeate the quantum continuum and space-time.

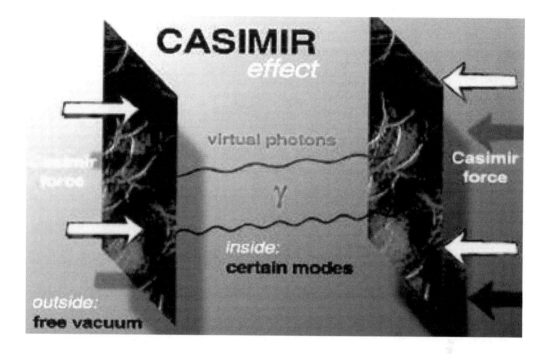

The extreme curvatures of space and the waxing and waning of holes could be explained as a wave function which dilates and contracts. In consequence, curvatures and valleys in space-time are constantly changing with distant areas brought together during contraction, and then later pushed apart as the crests and troughs flatten and even out, only to repeat the pattern of bunching together and creating tidal forces with crests and troughs, coupled with all manner of changes in frequency and oscillation. This expanding and contraction could even give rise to the illusion of an expanding universe and a billion years later a contracting universe.

Quantum mechanics allows for energy conditions to be violated "locally" at certain points (Fewster 2000; Ford & Roman 1995; Neumann 1937, 1955). Thus there are numerous regions in the quantum continuum which may be subject to negative energy fluxes. Likewise, these quantum inequalities can punch holes in space-time. Quantum inequalities are not theoretical but have been demonstrated in the electromagnetic field and the Dirac field--the quantum field for the electron (Fewster 2000; Ford & Roman 1995; Schrödinger & Dirac 1936).

Quantum Physics of Time Travel

Quantum inequalities in negative and positive energy always result in positive energy overcompensating for the presence of negative energy; that is, more positive energy must rush in than to where the negative energy was, more positive for less negative (Fewster 2000; Ford & Roman 1995). These inequalities create propulsion and yet another force straining against and creating holes in space-time.

The net effect of all these forces is to alter the "permeability" of these curvatures (or bilayers) and create "holes" allowing the blow out of negative energy or the temporary passage of positive energy from one side of the bilayer (space-time) to the other so as to reach a state of equilibrium; similar to processes are involved in diffusion and osmosis and the conductance of electric charges (Pollack & Stump, 2001; Slater & Frank, 2011; Wilf et al., 2007).

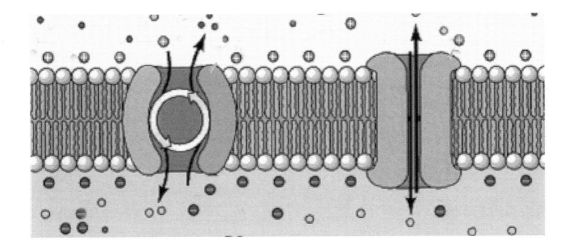

Negative Energy and Transmembrane Time Machines

If a transmembrane time machine were constructed so as to harness these imbalances, it could be actively carried from the hole's entry to its exit point at near light speed.

For example, imbalances between electromagnetic fields can create a high energy density on one side of a layer of space time, which in turn provides the energy for restoring equilibrium (cf Pollack & Stump, 2001; Slater & Frank, 2011). The creation of a "hole" or tunnel between two bilayers of space-time would therefore allow for both passive osmosis from high to lower pressure fields, and active transport to speed up the process, with the active energy provided by electromagnetism. The space-time traveler need only harness this energy or hitch his space-time machine to an electric field which is propagating through the vacuum between these layers.

For example, in biology, molecules can be transported by channel proteins or carrier proteins (Cooper 2009). Channel proteins cross various membranes in the energetically favorable direction, as determined by electrochemical gradients—a process known as passive transport. By contrast, carrier proteins are energized and serve as a vehicle and a mechanism which can use the energy for transporting molecules across a membrane.

If the time traveler was already at near light speed upon approaching the worm hole, he would also be molecular in size, if not smaller. If the time-space machine were designed to mimic a carrier protein, or to carry with it, like a booster rocket, a mechanism designed to mimic carrier proteins, he could be quickly transported from one end of the hole to the other.

Since mass can be converted into energy, energy into mass, these carrier proteins can also exchange energy or use these energy sources for the production of other forms of energy which can do work. For example, molecules can be transported in an energetically unfavorable direction across a membrane (e.g. against a concentration gradient) if their transport in that direction is coupled to ATP hydrolysis as a source of energy—a process called active transport (Cooper 2009). Thus, active transport can be used to go against the gradient, to transfer from areas of low to regions of high concentration.

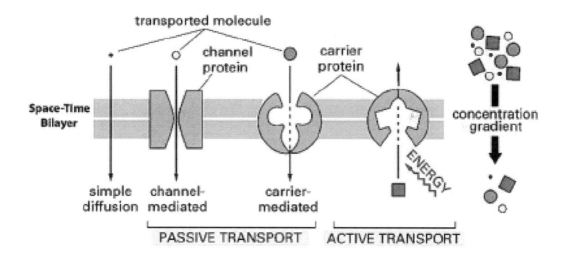

As pertaining to space-time travel, the time machine constructed to mimic the characteristics of a carrier protein could ride electromagnetic charges to go in any direction, even entering a hole which serves as an osmotic exit, and exiting the other end of the hole through which positive energy may be pouring in.

Using the analogy of a booster rocket, the space-time machine could consist of positively charged rockets and those with zero charge. For example, as in biology, under certain conditions, tiny pores, or "holes" in the bilayers of space-time may form which allow uncharged molecules and particles to pass between these curvatures, whereas charged molecules are blocked regardless of size (Cooper 2009).

Quantum Physics of Time Travel

A transmembrane space-time machine could act as a transporter by influencing the permeability of local regions within the bilayer thereby punching holes in space time and allowing passage between these layered curvatures. In biology, transmembrane proteins perform such functions between the bilayers of cells (Cooper 2009). These transporter transmembrane time machines could facilitate passage by generating gravity or energy or binding to positively charged objects which are actively carried through the bilayers and then to the otherside.

As in biology, transporters could also create conformational changes that open channels through which the time machine can be transported quickly through the hole and be released on the other side. In fact, he may be transported at light speed. As specified by James Clark Maxwell's equations (Pollack & Stump, 2001; Slater & Frank, 2011), when charges move they produce a twisting, or kink, in the electric field which propagates at the speed of light in the form of an electromagnetic wave, from one field to the other. Further, these waves are conductive and propagate because the changing electric field produces magnetic fields and vice versa. Thus, charged particles, differential positive vs negative energy densities, and the layers of space time which contain them, can interact with the charged particles on either side of the curvature of space time, via these fields which travel at the speed of light. A time-space machine constructed to ride an electromagnetic wave, since it will already be the size of a small particle if traveling at near light speed, would therefore arrive at the opposite end of the tunnel of time far into the future.

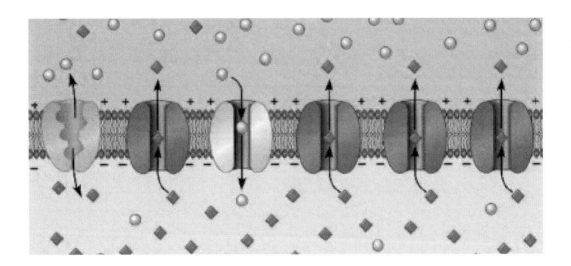

The Negative Energy Negative Mass Time Traveler

Be it through the action of transporters, concentration gradients, electromagnetic waves or other factors, these pores, holes, channels, and passageways would not remain permanently opened (Cooper 2009; Pollack & Stump, 2001; Slater & Frank, 2011). They may quickly close in response to local conditions on either side of the curvatures. The danger is that once the time traveler enters the hole, it may quickly evaporate and close up, trapping him inside forever.

Then there is the likelihood that the time traveler and her time machine will be stripped of positive energy when she enters the negatively charged vacuum between these folded up layers of space-time. She may emerge with negative energy and negative mass and then discover that she can't land her craft, as negative energy is repulsive.

For example, when journeying through a worm hole, the time traveler will likely have to pass through a negatively charged vacuum that forms between curved layers. As positive charged objects pass through the vacuum of a worm hole they would lose energy and have a negative charge when they emerge. By contrast, the mass or positive energy of a worm hole and surrounding vacuum state would increase because it is stripping away energy and mass from the object, in this case, the time machine. When the time traveler emerges from the other end of the hole, she may consist only of negative energy, have negative mass and still be atomized in size. She may also find it is impossible to land or get out of her time machine, for if she has negative mass and negative energy, she will be propelled away from all objects with positive energy.

If the gravitational mass of a particle is negative its inertial mass may also be negative. For a negative mass particle, although its force may be directed, for example, upward, it will accelerate downward such that the force and acceleration goes in opposite directions (whereas in a positive mass particle they are always the same). For example, if a pitcher threw a negative mass baseball, the batter would have to swing round and hit it in the direction it is going in order to make it go the opposite direction. If the batter swung a positively charged bat at a negatively charged baseball he would cause the ball to accelerate toward him and then over the plate and into the stands and seats behind him. This is because for negative particles force and acceleration are oppositely directed.

Regardless of if the world at the other end of the worm hole was negative or positive, the time traveler would be repulsed, and even sent at an accelerated speed in the direction she was already going and well beyond her destination point. Positive energy would propel the negative energy time machine to accelerate in the direction it is already going and would provide additional energy to accelerate its movement. Even if a positively charged object or pocket of positive energy were to directly collide with a negatively charged object, that object would instead accelerate to a greater velocity as if it was hit from behind.

If considered in terms of the 3 dimensions of space, the negative pressure would act in three directions, giving it a three fold effect and making the repulsive gravitational effect three time larger and creating an overall gravitational repulsion which could effect the 4th dimension of space-time. That is, after diving into a worm hole (or black hole) filled with negative energy, and upon being stripped of positive energy she would be propelled through the hole at superluminal speeds and then emerge with negative energy and negative mass at the other side and then be propelled to well beyond the speed of light in a negative direction, into the long ago. The Time Traveler, traveling at superluminal speeds, may also emerge from either end of the hole before or at the same time she enters. In fact, she may self-annihilate if there is a collision with her positively charged double which is heading for the future while she is heading from the future for the past. Duality is a natural consequence of traveling first to the future to get to the past. The time traveler will continually pass herself coming and going.

13: Black Holes And Gravitational Sling Shots

Due to the limitations of current technology one can accelerate to only a fraction of light speed. For example, one of the fastest vehicles on Earth is the Bugatti Veyron Super Sport's car, which has attained a velocity of 429.69 km/h (266.99 mph). Andy Green rode the ThrustSSC vehicle to a speed of 1228 km/h (763 mph). The New Horizons space craft shot into space in January 2006 and reached a velocity of 36,373 mph when its engines shut down. The Helios space probe satellite launched in the 1970s attained a velocity of 240,000 km/h (157,000 mph). However, the fastest manned vehicle was the Saturn V rocket which reached a velocity of 24,000 mph.

Since the speed of light is 186,282 miles per second (670,616,629 mph) the ability to fly even a few hours into the future will require new technologies, based perhaps, on matter / anti-matter engines or propulsion systems designed to harness the repulsion powers of negative energy. For example, particles with positive and those with negative energy are attracted to and repelled at the same time. When confined to within a closed space they will circle round each other, accelerating to greater speeds, although they will always maintain the same distance, thereby reducing the risk of mutual annihilation. A space-time machine with circular vacuum power sources containing pockets of positive and negative energy could potentially propel the time traveler to superluminal speeds.

Another method may be to locate one of the innumerable massive black holes which are believed to populate the spiral arms of the Milky Way Galaxy by the hundreds of millions (McClintock, 2004; Schödel, et al., 2006). The time traveler need not dive into the hole, but could utilize the surrounding vortex of space-time as a gravitational sling shot, thereby propelling the time machine toward light speed and into the future, and then to superluminal speeds into the past.

The Future Leads to the Past

To take a trip into one's personal past or to the past history of one's home planet, one can rely on memory, books, movies, imagination, or accelerate beyond light speed. Superluminal time travel is the only way to personally, physically, gain access to and directly visit the past and to go back in time.

Although it is possible to travel at a velocity slower and at a gravity lower than Earth, all that will happen is that clocks on Earth will slow down relative to the slow moving craft whereas the time traveler's clocks will run faster. In consequence, the time traveler will only discover that it takes more time to take a voyage to the future and that he will have aged more than those back on the home planet (the twin paradox in reverse). Likewise, it is not possible to go so slow as to reach a negative velocity and drift into the the past. She who wishes to travel in time will only fall further and further behind those speeding off toward the future. Such an endeavor will only make time pass more slowly for the intrepid time traveler. Nor is it possible to simply go in the opposite direction of light or the spin or orbit of Earth to travel to the past. It was demonstrated by Edward Morley in 1887 that despite the fact that Earth is accelerating and spinning in one direction, the velocity of light remained the same regardless of which direction it travels. The past is only accessible by traveling faster than light speed.

A seeming paradox of time travel into the past is that one must first journey to the future to reach the past; and this also calls for superluminal Lorentz transformations which many believe is not possible. In fact, the major objections to time travel into the past are based on Lorentz transformation equations and Einstein's special theory of relativity which decreed, by law, a cosmic speed limit and which erected a cosmic stop sign which proclaims: Nothing can travel faster than the speed of light. Einstein also proclaimed "God does not play dice." However, Einstein's cosmic speed limit is man-made and not dictated by a "god" whereas Einstein's theory of gravity allows for superluminal speeds.

As predicted by Einstein's theories of relativity, to exceed the speed of light one must first accelerate toward light speed and thus, to the future. For example, at 90% light speed the time traveler can reach the future in half the time its takes those back on Earth; one day in the time machine leading to 2.29 days in the future. At 99% light speed, one can leap 6 days into the future in just 26 hours. At 99.999999999999 the speed of light, one could leave Earth in the year 2050, and arrive in the year 2,2050 in 24 hours; two thousand years in a day. Time continues to shrink and the distance between the present and the future continues to contract with velocities approaching light speed, until at 100% light speed, time stops and one sits upon a event horizon with no future, no past, and an infinite "now." On one side of the horizon lies the future, on the other, the past.

At a speed of 99.99999999% c, one is hurtled far into the future. At light speed one arrives in the eternal present, the horizon of eternal "now." At 100.1% light speed the time traveler crosses over the horizon and continues into the past.

It is only upon accelerating toward light speed, which compresses time allowing the future to arrive more quickly, and then accelerating beyond light speed which triggers a time reversal; a consequence of the contraction of length, time-space, and time continuing in a negative direction. The time traveler must first journey to the future before continuing into the past.

Gravitational Sling Shots Into the Future And The Past

As of this writing, the only way to take a trip into one's personal past or to the past history of one's home planet, is to at accelerate beyond light speed. This may be accomplished using the gravity and energy of a massive black hole to propel the space-time machine to beyond light speed. Another alternative is diving into a black hole, at which point the time traveler will be accelerated to faster than light speed and then enter the mirror at the opposite end of the hole which leads to the long ago.

Space is permeated by holes: those smaller than a Planck Length; Worm holes of varying size which open up between folded layers of space-time; Holes created by the compressed mass of four or more sun-like stars, millions of which litter the arms of the Milky Way Galaxy; Super-massive black holes with the concentrated gravity-mass of millions of billions of stars and which sit at the center of spiral galaxies (Blanford 1999; Melia, 2003a,b; Jones et al., 2004; Ruffini & Wheeler 1971). Super massive black holes have such incredibly powerful gravitational tidal forces that they suck entire stars and their planets into its depths, never to be seen again (Giess, et al., 2010; Melia, 2003a,b; Merloni & Heinz, 2008).

The gravity of massive black holes sprinkled throughout the outer arms of the galaxy, and those at the center of galaxies, offer the time traveler a means of accelerating to near light speed and into the future, and beyond light speed into the past. A black hole can be employed as a gravitational slingshot.

Essentially, the gravity of a black hole can be borrowed to increase the space-time machine's velocity. The space-time machine would fly toward a spinning black hole which would cause the craft to accelerate to near light speed (Penrose 1969).

While remaining a safe distance from the hole's event horizon, the time traveler could then ride piggy back on the movement and energy of the spin to be flung around the exterior of the hole at a speed equivalent to the craft's velocity plus the near light speed velocity of the outer rim of the vortex.

General relativity predicts that a rotating black hole will drag space-time around it and eventually into the hole. Surrounding the "event horizon" of a black hole is a vortex of space-time referred to as the ergosphere (Bethe et al. 2003; Taylor & Wheeler 2000). Objects, particles, space-time machines, theoretically, would be able to escape from the ergosphere. Through the Penrose process (1969), objects can emerge from the ergosphere with more energy than they entered. The energy and accelerated velocity gained by the space-time machine would be equal in magnitude to that lost by black hole; which, given the relative differences in size would be negligible insofar as these stellar objects are concerned. However, the combined velocities could propel the space-time machine to well beyond the cosmic speed limit.

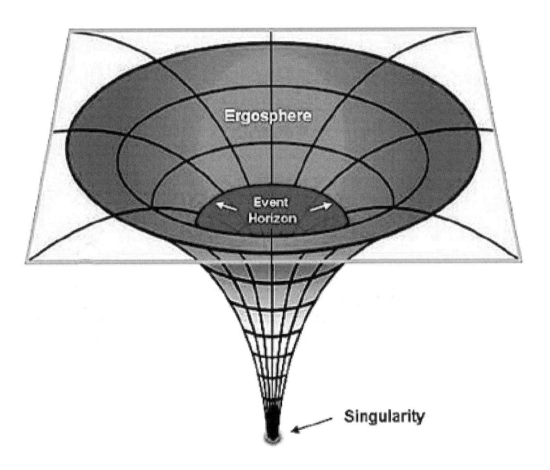

For example, if a time traveler were standing on a railroad track with a tennis ball in hand, and tossed the ball at 25 mph at a train approaching at 100 mph, passengers in the train would see the ball approaching at 125 mph, and then bouncing off the train at nearly the same speed. Likewise, if a professional baseball pitcher was riding in the front end of the lead locomotive racing along the tracks at 100 mph, and he threw a base ball at 100 mph to someone standing in front of the train, the initial speed of the ball would be 200 mph relative to the ground. The additional energy and acceleration in both instances are borrowed from the train.

General relativity predicts that any rotating mass will "drag" and pull space-time which begins to circle around it. Objects closest to the black hole have a significantly higher velocity that those further away. Therefore, using the Penrose process the time-space machine can ride on the spin of a spinning black hole, and by employing a parabola trajectory, circle round at an accelerated rate of speed and leave the vicinity in the opposite direction at beyond light speed. The ergosphere of the black holes becomes a gravitational sling shot.

By using a parabola trajectory the time traveler could avoid the Schwarzschild radius and event horizon where the gravity and curvature of space is so severe that the space-time machine would be unable to escape. The point of no-return is the event horizon (Bo & Wen-Biao, 2010; Melia, 2003b; Hawking 1990; Thorn 1994; Thakur, 1998). Just as a ship can't be seen after it crosses over the horizon on Earth, everything that passes over the "event" horizon of a black hole can't be seen--though it may be remembered.

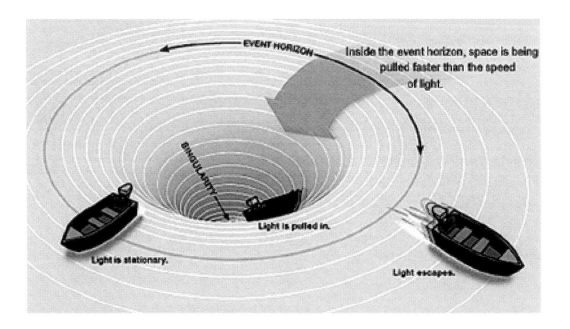

The size of the event horizon is given by the "Schwarzschild radius" (Brill et al. 2003) and anything which falls through the radius, and over the horizon, is cut off from the outside universe (Bethe et al. 2003; McClintock 2004). What happens next is anyone's guess. Possibly everything decays and is torn apart due to the incredible tidal forces, with even atoms becoming ripped apart and crushed to near infinite density; unless what falls in is already atom-in-size or smaller than a Planck length--a consequence of length contraction as velocities approach light speed.

Tidal forces are believed to be smaller at the event horizon, even less so at the ergosphere, and to decrease as the size of the hole increases (Bethe et al. 2003; Brill et al. 2003; McClintock 2004), Therefore, the bigger the hole, the more likely someone can survive, especially if the time machine has shrunk to the size of a Planck length, at which point it may also blow a hole in space time and be propelled to the other side.

The best way to avoid the many perils which may await the intrepid time-traveler within the darkening depths of a black hole, is to use its gravity to increase the speed of the time machine and then go around it.

Relativity, Space Time...

A parabola orbital path would enable the time traveler to avoid the event horizon and borrow the speed, gravity and the spin velocity of the black hole which would accelerate the time-space machine as it approaches the hole. Combined with the near light speed of the outer vortex of the hole and the craft's near light speed velocity, the time machine would be flung into the past at superluminal speeds.

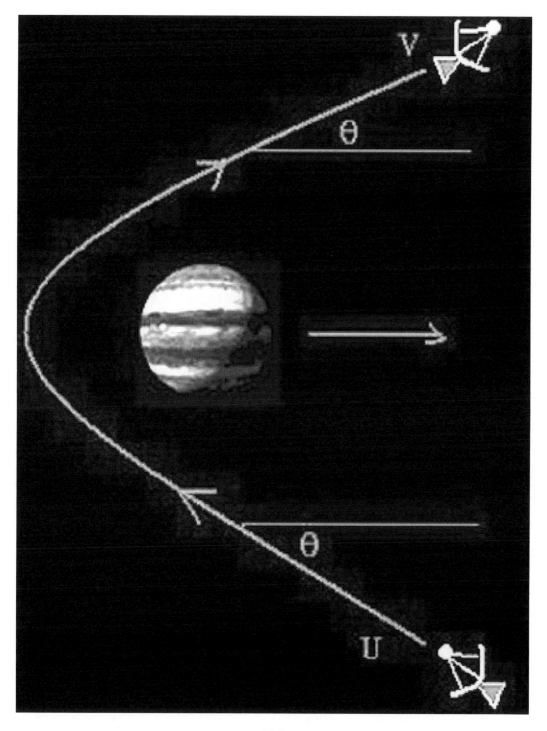

Gravitational slingshots, also known as gravity assist maneuvers, have been employed to increase velocity by numerous interplanetary missions beginning with the Mariner 10 as well as the Voyager probes which used the combined gravity of Jupiter, Saturn, Uranus and Neptune which were closely aligned at the time. An alien space-craft approaching from another solar system could use our sun or this entire solar system for the same purposes.

Thus, a spinning black hole could also serve as a slingshot; though if superluminal speeds can be achieved is open to debate and would be dependent on numerous variables. For example, if the time traveler misjudges the gravitational grip of the black hole and is caught up in the vortex of surrounding space, she will not only accelerate to near light speed but she may be pulled to the hole's event horizon where she will spin round and round the event horizon at the speed of light. It is only upon falling, or diving into the hole that she may exceed light speed and be hurtled toward the past.

Hence, aiming directly for the heart of a black hole is yet another means of achieving superluminal speeds. This course of action, however, also poses many perils including the possibility the Time Traveler will emerge no larger than an atom, and with negative mass and negative energy.

14: The Time Traveler in Miniature: Negative Mass and Energy

Upon accelerating toward light speed the time-space traveler and her time machine will gain incredible mass and energy and shrink and contract in size. For example, if a 1000 foot in length time-space ship attained a velocity of 60% light speed, it would contract by 20%, to 800 feet. The amount of length contraction can be calculated and determined by the Lorentz Transforms (Einstein et al. 1923).

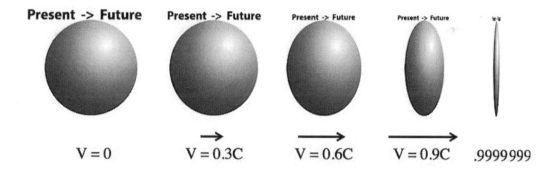

If her time machine has spin, rotation, and acceleration the phenomenon of length contraction will be equalized and she and her vehicle will shrink to molecular and then atom sized; otherwise she will be spaghettified and be all height yet hair-like thin as length contraction is always in the direction of motion. Moreover, although growing smaller in size she will be gaining mass as predicted by general relativity (Einstein 1915a,b, 1961).

According to this equation: $E = mc^2$, where E is energy, m is mass and c is the speed of light, mass and energy are the same physical entity and can be changed into each other. Because of this equivalence, the energy an object has due to its motion will increase its mass. In other words, the faster an object moves, the greater its mass which at the same time is contracting. This only becomes noticeable when an object moves at great speeds. If it moves at 10% the speed of light, for example, its mass will only be 0.5% more than normal. But if it moves at 90% the speed of light, its mass will double and it will also contract by 50%. As an object approaches the speed of light, its mass rises precipitously as it shrinks to nearly nothing in size.

Upon reaching a velocity of 186,000 miles per second, time stops and the time traveler will experience an eternal present which is frozen in place. The time traveler will also have shrunk to a size perhaps smaller than a Planck Length. It has been theorized, based on Einstein's theories of relativity, that at this speed she and her time machine will have infinite mass and will have disappeared into nothingness. However, this prediction may represent the limitations of Newtonian and Einsteinian physics, the applicability of which diminishes and loses all meaning when applied to space smaller than a Planck length, the smallest unit of measurement. Space-time, within the Planck scale, is subject to extreme uncontrollable quantum fluctuations, as it is continually being bent, folded, crumpled, and torn apart by powerful gravitational forces; and the laws of physics break down (Bruno, et al., 2001).

Although relativity says mass becomes infinite at or beyond the speed of light, quantum mechanics does not support this prediction. Infinity is possible only in an infinite universe (Joseph 2010bc). In a finite (vs infinite) universe, the concepts of infinite mass and infinite energy may well be mathematical anomalies which represent not the impossibility of accelerating above light speed and then back below the speed of light, but the limitations and breakdown of those theories which are no longer applicable when applied to events taking place at these velocities and at these small sizes. Most physicists would agree that Newton's laws of motion are suitable for macro-objects whereas quantum mechanics must be applied for micro-structures if there is any hope of obtaining measurements and making predictions which might agree with experimental observation (Heisenberg 1927, 1958). However, at an atomic scale the energy and speed of a particle is uncertain and the smaller the size the greater the uncertainty and the greater the fluctuations. In fact, at the Planck length, the Uncertainty Principle and quantum indeterminacy becomes virtually absolute (Eisberg & Resnick 1985; Heisenberg 1927, 1958; Smolin, 2002).

If particles, objects, or time-machines contract to a size smaller than a Planck Length as they near and then reach light speed, they have so much energy they may instead create their own hole in space-time as predicted by quantum mechanics, and then be expelled at superluminal velocities with negative energy and negative mass; with positive energy acting as propulsion and negative energy as repulsive, the close proximity of which increases escape velocity. The same consequence may befall all those who dive into a super-massive black hole with a negative energy density.

Newton's and Einstein's theories of gravity actually predict that if increasing mass is shrunk to a subatomic space, its gravity will become increasingly powerful, whereas, as predicted by quantum physics if the object shrinks to a size smaller than Planck length, it will have so much Planck energy (10^{19} GeV), that coupled with its powerful gravitational forces (Eisberg & Resnick 1985; Smolin, 2002) that it will both collapse and blow a hole in space time about the size of a Planck length.

Relativity, Space Time...

Figure: Shrinking to a Planck Length and creating a hole in space-time as a photon circles by.

As to the possibility of "infinite mass" photons travel at light speed and photons have zero mass, not infinite mass. Likewise, it can be predicted that a time machine upon reaching light speed, may also have zero mass, a consequence, in part of its release of energy when it shrinks to a size smaller than a Planck length and blows a hole through space-time (Joseph 2010b).

Infinite mass and infinite energy are possible only in an infinite universe (Joseph 2010c) and thus predictions of infinite energy and invite mass in a finite universe should be viewed as mathematical anomalies. In a finite (Big Bang) universe, and as predicted by quantum mechanics, it is more likely that at light speed the time traveler will have shrunk to a size smaller than a Planck length. Because so much energy will be released, instead of infinite mass, she would have no mass or negative mass and negative energy when she emerges from the hole, be it a worm hole or super massive black hole.

Worm holes, for example, tunnel through a negatively charged vacuum between the layers of space time which would strip away positive energy to equalize energy gradients, such that the Time traveler would have have a negative charge when they emerge. The same stripping away of positively energy may also take place when passing through a super massive black holes. It is precisely because these holes may be filled with a negative energy vacuum that the Time Traveler may be stripped of positive energy and mass while being propelled to well beyond the speed of light into the future, the past, or a mirror universe.

Upon entering this mirror universe, she not only accelerates into the past but she may consist only of negative mass and energy; a phenomenon which also allows for the creation of duality without violating the laws of mass and energy conservation. There is no violation because she will consist of negative mass and energy and thus no extra mass or energy is introduced and as the negative equalizes the positive energy / mass of the time traveler who already exists in the past.

Duality is a natural consequence of traveling first to the future to get to the past. The time traveler will continually pass herself coming and going.

Moreover, due to length contraction, upon emerging from the hole, be it a supermassive black hole, worm holes, or hole the size of Planck Length, she may be no larger than the smallest particle or emerge as electromagnetic energy which radiates out from the hole into space (Giddings, 1995; Hawking, 2005, 2014; Preskill 1994; Russell and Fender, 2010), possibly speeding off at superluminal speeds into the past.

Coupled with negative mass and negative energy, the time traveler upon emerging from the future (a consequence of accelerating toward light speed) and then speeding from the future into the past, may share characteristics with superluminal (hypothetical) particles such as "tachyons" (Bilaniuk & Sudarshan 1969; Chodos 2002; Feinberg, 1967; Sen 2002). The hypothetical tachyons are believed to always travel at superluminal velocities. Tachyons, which may have negative energy and negative mass, are the true time travelers, forever journeying from the future to the past.

15: Tachyons, Negative Energy, The Circle of Time: From the Future to the Past

It is well established that various particles have a velocity close to or at the speed of light (Houellebecq 2001). Many of these high speed particles were hypothetical until their existence was verified experimentally. Some particles, such as positrons, and hypothetical "tachyons" are, or were believed to travel faster than light (Bilaniuk & Sudarshan 1969; Chodos 2002; Feinberg, 1967; Feynman 1949; Sen 2002) whereas others, such as photons and electromagnetic waves travel at light speed. Superluminal tachyons, however, if they exist, may have negative energy and negative mass (Chodos 2002; Feinberg, 1967).

Electromagnetic waves are a fundamental quality of matter and are subject to the effects of gravity including galactic lensing (Slater & Frank, 2011; van der Wel 2013). When the electromagnetic force is stripped of its particle, it has no mass, and this is what is believed to occur when particles, or time machines enter a black hole (Everett & Roman 2012); what emerges has no mass, or even negative mass and negative energy and may journey at faster than light speeds.

Wheeler and Feynman argued in 1945 that electromagnetic waves emitted by an electron proceed into the future and the past (Wheeler & Feynman 1945, 1949). When these waves collide with waves in the future they send waves back in time and further into the future due to the collision. Those sent to the past can also collide with those in the past sending them again in opposite directions, into the future and further into the past. Depending on if these waves collide crest to trough their energy levels may double, but if they collide crest to crest (or trough to trough) they cancel each other out; a phenomenon referred to, respectively, as constructive and destructive interference.

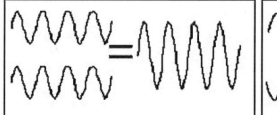
Constructive Interference

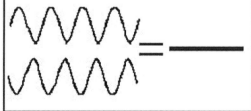
Destructive Interference

Quantum Physics of Time Travel

Space-Like and Time-Like Separations

Some hypothetical particles, such as the "tachyon" are believed to be time-independent and to travel faster than light speed (Bilaniuk & Sudarshan 1969; Chodos 2002; Feinberg, 1967; Sen 2002); meaning that these particles are constantly arriving in the present from the future and continue their high speed journey into the distant past.

In contrast to slower than light particles which have "time-like four-momentum" these hypothetical superluminal particles have "space-like four-momentum." For example, if two events have a greater separation in time than in space, they have a time-like separation which is indicated by a negative (minus) sign. If the sum is positive, the two events have a space-like separation which is greater than their separation in time. If the result is 0, then the two events have a light-like separation and are connected only by a beam of light.

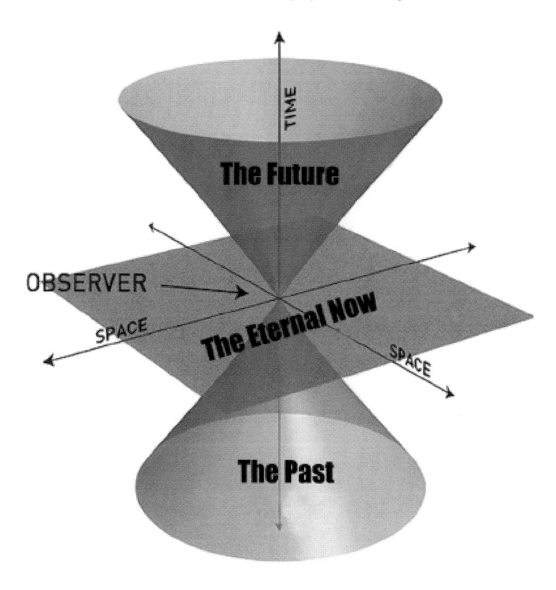

Relativity, Space Time...

For example, if a football player throws a pass and it is caught by a retriever 50 feet distant after an elapse of 5 seconds (5000000000 nanoseconds), then the separation in space is 50 and the separation in time is 5. By calculating the square of the separation in space minus the square of the separation in time, it can then be determined if the separation of the two events are space-like or time-like:

50^2 (feet) - 5000000000^2(nanoseconds) = 2,500 - 500000000000 = -499999998500

In this instance, since the results are a negative, the two events have a greater separation in time than in space.

The star Proxima Centauri is 4.2 light years from Earth. If the time traveler was required to visit that star 10 years from today, then they would be separated by 4.2 light years in space and 10 years in time. Thus it would be 4.2^2(space) - 10^2 (time) = 17.64 - 100 = -82.38 meaning that the two events are separated in time and have a -82.38 time-like separation from now. However, if the time traveler had to arrive on Proxima Centauri in 3 years, then it would have a space-like separation and it would be impossible for him to get there is 3 years since there is not enough time; unless he were to exceed the speed of light or travel via dreamtime.

In relativity, space and time are combined as a single continuum within a fourth dimension: space-time (Einstein 1961). In space-time, the separation between two events is measured by the invariant interval between the two events, which takes into account not only the spatial separation between the events, but also their temporal separation. For two events separated by a time-like interval, enough time passes between them that there could be a cause–effect relationship between the two events.

In space-time, a coordinate grid that spans the 3+1 dimensions locates events (rather than just points in space) whereas the other 3 dimensions (considered separately) locate a point and location in a certain defined "space." The spatial location of an event is designated by three coordinates, X, Y, Z, whereas a fourth coordinate is based on time; all of which constitute "frames of reference." Together the 3+1 dimensions locates events and when and where they took place. Therefore, in relativistic contexts time remains entangled with the other three dimensions of space (length, width, height) as well as with the space-time quantum continuum. The world-line of time can extend from the past to the future and from the future to the past. In qunatum theory, and as based on evidence of entanglement, the world-line of time can become space-like and this would permit information to flow from the future to the past, and at superluminal speeds.

Tachyons

Tachyons are believed to have worldlines which are space-like and not time-like such that the temporal order of events would not be the same in all inertial frames (Bilaniuk & Sudarshan 1969; Chodos 2002; Feinberg, 1967; Gibbons, 2002; Sen 2002); meaning cause and effect would be reversed or abolished. Tachyons, because they travel from the future to the past violate causality.

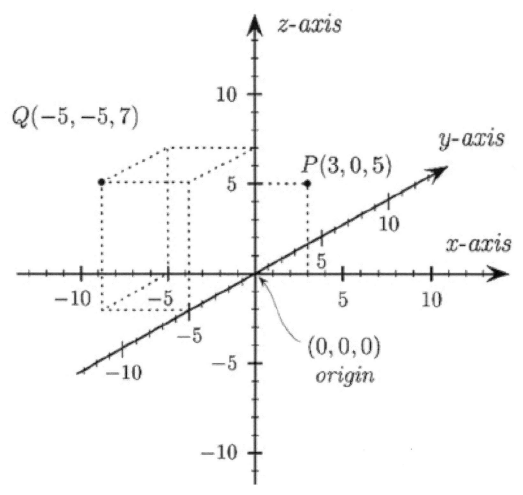

Figure: the spatial location of events and time

    The existence of a tachyon particle was first proposed by Gerald Feinberg in 1967. According to Feinberg's theories and calculations, a tachyon could be similar to a "quanta" of a quantum field but with negatively squared mass; that is, it would have no mass or anti-mass. Its energy, nevertheless, would be real. However, whereas objects traveling toward light speed gain energy and mass, if the tachyon increases it superluminal speed it loses energy. If it accelerates to 200% light speed its energy diminishes to zero. Therefore, whereas objects are believed to reach "infinite velocity" upon attaining light speed, the tachyon can only reach infinite velocity upon accelerating to speeds twice that of the speed of light, at which point there would be another time reversal and the tachyon would journey from the past into the future.

Relativity, Space Time...

The existence of "tachyon" like particles has been rejected because they would violate the laws of causality and Einstein's theory of special relativity (Aharonov et al. 1969). Feinberg (1967), however, determined that special relativity did not prohibit faster than light travel so long as the object had always maintained superluminal velocities and had never had a velocity below the speed of light. According to special relativity (but not general relativity or quantum mechanics) the acceleration of matter to beyond light speed could cause the energy of this mass to becomes infinite and the Lorentz transformations would then have no meaning. However, if superluminal velocities are the norm for these faster than light particles, then there would be no need to break the cosmic speed limit except at 200% light speed. By the same token, these particle would never be able to reduce velocity to below light speed. Therefore, if these and other particles are traveling beyond the velocity of light they may have always journeyed at superluminal speeds and never had a velocity below light speed.

Others have argued that particles with negative mass cannot travel faster than light and would have negative energy and become unstable and undergo condensation (Ahraonov et al. 1969). These arguments were countered by Chodos (1985) who proposed that neutrinos can behave like tachyons and travel at superluminal speeds. By violating Lorentz invariance, neutrinos and other particles would undergo Lorentz-violating oscillations and travel faster than light while maintaining high energy levels. However, over time superluminal neutrinos would also lose energy, probably as Cherenkov radiation (Bock 1998).

Although there have been numerous proposals and arguments to and fro it appears that objects or particles would lose mass and energy upon reaching superluminal speeds and those which travel faster than light would have negative mass and negative energy. Further, it could be said that tachyon-like particles once they accelerate to superluminal velocities, or, if they have always journeyed at faster than light speeds, may be unable to slow down to a velocity less than the speed of light. It has also been theorized that tachyons must maintain a constant speed, for if the Tachyon were to accelerate and increase velocity it loses energy which becomes zero if the speed reaches infinite velocity; i.e. 200% light speed; a velocity which triggers a time reversal with negative contraction imploding and continuing in a positive direction back toward the future from the past.

According to some theories, if a tachyon were to slow toward light speed, the energy of a tachyon would increase and would becomes infinite as its velocity equals the speed of light; and the same would be true at 200% the speed of light. This is a mirror image of what is theorized to occur when particles or objects reach light speed as they are also supposed to gain infinite mass and infinite energy. Thus particles which always travel above and those which always travel below light speed are mirror images of each other and may have the same barriers and non-traversable event horizon, with the past on one side and the future on the other.

Quantum Physics of Time Travel

Duality

As implied by the Lorentz transformation (Einstein et al 1923), a tachyon would always have negative energy. The Lorentz transformation indicates that the sign of a particle's energy is the same in all inertial frames, just as the sign of the temporal order of two points on the world line remain the same. All observers will see that the particle has positive or negative energy, though they may disagree on how much energy it has. However, if the particle has positive energy according to one observer, and negative according to another, then the observers, or the particles, are occupying different inertial frames (e.g. one in the present the other in the future/past; and this again implies duality.

If tachyons or other objects did not have the same energy sign in all inertial frames then perhaps they are looping in and out of the past and future, becoming positive when below the speed of light and negative above it as predicted by superluminal Lorentz transformations (Everett & Roman 2012). Because they would have positive energy when heading toward the future and negative energy when traveling into the past they could both exist even in the same inertial frame.

The dichotomy between positive vs negative energy and mass implies duality; the tachyon or time machine which voyages beyond light speed is the antithesis of the tachyon or time machine at a velocity below light speed. For example, the tachyon below light speed could be considered an anti-tachyon. The antiparticle of a tachyon would be a positive energy tachyon which is traveling forward in time. The negative energy tachyon would be coming from future heading into the past. As such, they would seem to be continually circling around each other from the perspective of an observer, one coming the other going in parallel continuously in every moment of time until one or both loses energy or arrives at the point when it enters the past from the future, or comes to the end of its journey into the past, its negative mass and negative energy spent.

In fact, since, according to Einstein (1955), the distinctions between past and future are an illusion, and as multiple futures and pasts exist simultaneously, overlapping in different regions of space time, the positive and negative tachyons could not only exist in parallel, but would continually be chasing and escaping from each other, with the positive energy tachyon showing attraction and the negative repulsion and with both maintaining the same distance from each other. Positive and negative tachyons, therefore, would create a circle of time.

Time can be experienced, and therefore time must have energy and potential mass, or a particle wave duality. "Tachyon" like positive vs negative particles would also have a wave function. Tachyon's might also be accompanied by gravitational waves which radiate in front of it or behind it; causing the tachyon to lose energy and to accelerate to even greater speeds--such that as it journeys into the past it would speed up, until losing all negative energy. However, if this would occur only at 100% or 200% light speed, and the loss of negative energy would be replaced by positive energy; and the cycle continues.

Relativity, Space Time...

Electrons, Positrons, Tachyons and the Circle of Time

Einstein's field theories predict the curvature of space-time, such that the universe and time circles back upon itself (Gödel 1949a,b). The future leads to the past; a realization which greatly troubled Einstein. The existence of particles which travel from the future to the past are a logical extension of Einstein's theories and field equations (Gödel 1949a,b). Time is a circle which may be orbited by positive and negatively charged particles. If correct, then these negative and positively charged particles would also create a neutral state of equilibrium (Feynman 2011; Pollack & Stump, 2001; Slater & Frank, 2011; Wheeler & Feynman 1949); much like the positively charged nucleus of an atom counters the negative charge of an electron--the amount of positive charge determining the number of electrons.

Electrons may also circle in and out of time and changing charges as they do so, with negatively charged electrons directed toward the past and positively charged electrons, referred to as "positrons" directed toward the future. John Wheeler proposed that all electrons and positrons (the antiparticle to the electron) have identical mass but opposite charges (see also Feynman 1949). According to Wheeler (2010; Wheeler & Feynman 1949), all electrons in the universe zig zag backward and forward in time, and when zigging backward it is an electron and when zagging forward it is a positron. And when zigging and zagging they interact as an electron-positron pair, moving in and out of the past and future. Richard Feynman (2011) incorporated these ideas in his formulations for quantum electrodynamics which earned him a Nobel Prize. However, these pairs are not necessarily being created or annihilated; though annihilation could be predicted if they were to come in contact. Rather, like the positive and negative charged tachyon, they chase each other in a circle of time.

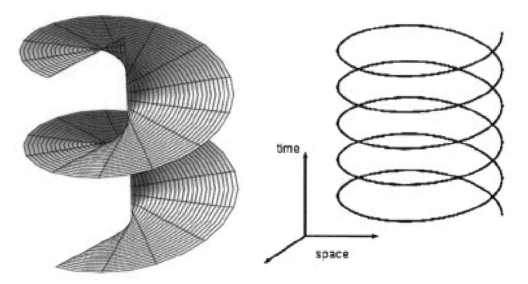

If we were building an atom-of-time, then it could be proposed that positively charged positrons and negatively charged tachyons circle toward the past, maintaining an equilibrium of charges in the past, whereas negatively charged electrons and positively charged tachyons do the same in the future; with all four circling around each other, in and out of the future and the past and without violating the laws of conservation of energy and mass.

Therefore, if a negative energy negative mass tachyon and a positron traveling from the future to the past was able to circle round and go from the past back toward the future, the tachyon would become a positive energy anti-tachyon and the positron a negatively charged electron. If time is a circle, the positron/electron and tachyon/anti-tachyon may circle from the future to the past to the future and back again; as if time was composed of particles which orbit the nucleus of "eternal now." Positrons, electrons, and tachyons would therefore provide time with the energy and atomic structure to emerge from the quantum continuum and be perceived as something real.

The Circle of Time

That time exists there can be no doubt. Time is experienced, is perceived, has length, duration and can be precisely measured. Thus, time must have substance and a particle wave duality. Although Einstein placed time in the 4th dimension along with space, he never told us what time is made of.

If the universe, and time, are curved and circles back, then what provides time with the energy to perpetually circle and to be perceived? What would prevent negative mass / energy and positive mass / energy from self-annihilating if they came into contact? The answer to all these questions may be repulsion vs attraction and negative vs positive energy and mass.

For example, if a negative object and a positive object of the same size came into contact, the negative would be repelled away and the positive would accelerate toward--a push pull scenario which could result in the negative and positive objects circling round and round each other as they are attracted and repelled at the same time--like a very bad romantic relationship.

An object with negative energy falls down just like an object with positive energy. However, if a negative particle swerved near a planet, the gravitational effect would be repulsive and it would be pushed away. Negative mass is repelled by positive mass and vice versa. If both were negative, they would also be repelled and this is because the two minus signs (-m and -m) cancel each other out.

Positive energy would propel the negative energy object to accelerate in the direction it is already going. If the universe and time are curved and lead back to their starting point in a circle, then the positive (heading toward the future) and the negative (head from the future to the past) would also circle round each other, with the future leading to and following into the past. If the positive particle actually caught up with and bumped into the negative particle, such as might be expected at the event horizon, the positive would force the negative to speed up in the negative direction it is already going.

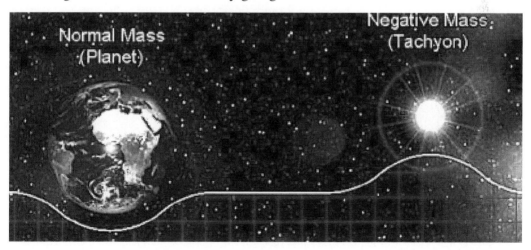

Moreover, although the the positive and the negative particles might maintain the same distance from each other, they would accelerate to greater and greater speeds--and this is because both have acceleration; despite the fact that this seemingly violates the conservation laws of momentum and energy which requires that they remain constant.

For example, the positively charged anti-tachyon would accelerate toward the future coming closer and closer to light speed, and upon crossing the event horizon separating future from past, would lose positive energy and attain negative mass and then accelerate backwards into the past at superluminal values. However, they would also chase one another, such that both increase in speed; the positively charged tachyon toward the velocity of light, and negatively charged tachyon to twice the speed of light; or, in a mirror universe where all is reversed, the negative would be forced to below the speed of light. That is, as the positive speeds up, the negative, going in a negative direction, might slow down, with both exchanging energy at the event horizon of "eternal now." Alternatively, the tachyon may accelerate until reaching twice the speed of light thereby losing negative energy and gaining positive energy as the contraction of time-space implodes and collapses in a positive direction.

At this juncture, we can only theorize and hypothesize: negative energy tachyons become positive energy anti-tachyons and positrons become electrons; and the circle of time continues to circle around with the positive chasing the negative which is chasing the positive, like the hands of a clock.

The Positive and Negative Time Traveler

It has been predicted that a Time Traveler, upon accelerating beyond light speed would lose all positive energy and mass and consist only of negative energy and mass. This is a consequence of length contraction which continues in a negative direction after exceeding light speed, and the result of entering a black hole in space-time. If the time machine had positive energy which was stripped away by negative energy, the time traveler would still be propelled through the hole and then emerge with negative energy and negative mass at the other side. The time traveler would then continue in a negative direction, into the long ago.

Like a tachyon and an anti-tachyon, or an electron and a positron, a time traveler would also become a duality; a positive mass/energy time traveler heading towards the future and a negative energy/mass time traveler heading from the future into the past. However, if the positive and negative come in contact, they may self-annihilate and one or both must disappear, leaving behind only the mass/energy of one. Again referring to the theories of John Wheeler and Feynman (1945; Feynman 1949, 2011; Wheeler 2010), this could explain why there are more electrons than positrons in the universe. However, this also suggests that a Time Traveler faces extreme danger in traveling to the past. If he is not forced to travel in an endless circle from the future to the past and from the past to the future with the positive chasing the negative, he may come into contact with his double that is heading toward the future or coming from the future into the past, resulting in self-annihilation.

16: Duality: The Past And Future In Parallel

Just as an object (unless destroyed) is continually visible in the present and remains visible an hour later, the next day, the next week, and so on, any particle or object heading for the future remains continually visible to observers external to the time machine; and this is because the present also advances to that future. Likewise, any object which turns around and heads for the past, would be continually visible to those in all past moments leading from the present to the past. That same object or time machine sent into the past would also be visible to observers as that past progresses to the present-moment with the object/time machine was sent into the past; like watching a movie in reverse.

H.G. Wells' Time Traveler and his time machine began the voyage through time in his laboratory. Although he could see events, people, and even a snail whiz by: "The laboratory got hazy and went dark. Mrs. Watchett came in and walked, apparently without seeing me, towards the garden door. I suppose it took her a minute or so to traverse the place, but to me she seemed to shoot across the room like a rocket."

A time machine, however, would not disappear but would be continuously visible as it continuously travels through time. Whereas Mrs. Watchet, the housekeeper, from the perspective of the Time Traveler, would seem to race across the room, from Mrs. Watchet's perspective the Time Traveler would appear to be shrinking in size and to be frozen in time or moving exceedingly slowly and perhaps enveloped in a red and blue shifted haze of light; a consequence of continually arriving in the future-present (blue shift) and continually traveling through the present to the future (red shift) or to the past-present (blue shift) and from the past-present to the past (red shift). Light becomes red shifted as it accelerates away from an observer, and blue shifted as it approaches.

A time machine would not disappear even if it exceeded the speed of light and then journeyed into the past. It would be visible to those in every moment of the past as the time traveler travels backward through it and as they travel forwards through it: like the same movie played on two screens side by side, with one movie playing from the end to the beginning and the other from the beginning to the end but with the characters in the movie able to see the movie showing beside them.

In the opening chapter of Well's story the Time Traveler ushers his friends and colleagues into his study where they sit down in his den. The Time Traveler describes his concept of time travel and a model of his time machine which he promises to demonstrate:

"The Time Traveller smiled round at us...and with his hands deep in his trousers pockets, he walked slowly out of the room, and we heard his slippers shuffling down the long passage to his laboratory.... The Time Traveller came back...The thing the Time Traveller held in his hand was a glittering metallic framework, scarcely larger than a small clock, and very delicately made. There was ivory in it, and some transparent crystalline substance...On this table he placed the mechanism....Then the Time Traveller put forth his finger towards the lever....We all saw the lever turn. I am absolutely certain there was no trickery. There was a breath of wind, and the lamp flame jumped. One of the candles on the mantel was blown out, and the little machine suddenly swung round, became indistinct, was seen as a ghost for a second perhaps, as an eddy of faintly glittering brass and ivory; and it was gone - vanished! Save for the lamp the table was bare" (H. G. Wells, The Time Machine).

If it had gone into the past, and unless the model of the time machine accelerated into space, then the guests would have seen it sitting on the table as they entered the room; it would have been ever present in the eternal now even as it was traveling backward through time. And the same would be true if it headed into the future. This is because time travel is not instantaneous, but more like running in place with space-time contracting and the future coming closer to the present. Even if the model of the time machine was accelerating toward light speed in orbit around Earth, it would be continuously visible, just as the International Space Station is visible even though it is traveling just a few clicks of the clock into the future at a rate slightly faster than those back on Earth (Lu 2000).

Since time travel is not instantaneous, the time machine does not suddenly leap across time, but instead journeys across time at a different velocity than observers outside the time machine.

Therefore, even if traveling beyond the speed of light, which is the only way a time machine can reach the past, the toy model of the time machine would have still been visible from the moment the Time Traveler's guests entered the room and sat down, to the moment the model was retrieved from the laboratory and sent into the past. Only then would it disappear. However, at the moment he brought into into the room, until the moment it disappeared, the Time Traveler's guests would have seen two model time machines.

Duality and Multiplicity

From the moment the Time Traveler's guests entered his house and stepped into his den, they would have seen the model time machine as it traveled into the past, occupying all their present moments going forward in time as it did so. However, since it was traveling faster than light speed, its reflected light would trail behind it, and at most, what they might see would be a ghost-like

Relativity, Space Time...

contraction of space which would also be enveloped in red shifts and blue shifts on the spot on the table top where the model sat as it journeyed into the past. And they would continue to see these ghostly flashes of lights even as the Time Traveler left for his laboratory to retrieve the model he set before them; at which point, his guests would see two models, one going backwards in time and continually passing through all present moments, and the other in the time traveler's hands as he brought it into the room and set it on the table.

In fact, as the model time machine was heading into the past, it was also still sitting in the laboratory.

Therefore they would would be two models, until the moment it was sent into the past and then both would disappear from all future present moments as it was heading backward in time. That is, they would see a model sent into the past sitting on the table from the moment they entered the room and the one that had not yet been sent into the past which the Time Traveler retrieved from his laboratory. And yet, here is a paradox which also seems to violate he laws of conservation of energy and mass: A journey into the long ago creates a duality which may become a multiplicity.

For example, as the time traveler accelerates to superluminal velocities and speeds into the past, he will be retracing his steps and may encounter himself in the past on his way to that moment when he left for the past. Likewise, as he accelerates toward light speed and into the future, he will see himself coming back from the future and heading again to the past. He will be continually coming and going, sharing the same "eternal now" in parallel, like the same movie running side by side on adjacent screens, but with one going backwards and the other forwards.

As to multiplicity, the guests of H.G. Well's time traveler would see two models on the table the one sent into the past, the one about to be sent into the past. However, at the moment it is sent into the past, two models would be sent into the past. Therefore, they would have seen two models when they entered the room. But as the process repeats, then there would have been two models going backwards in time and a third model that he brings from his room and sets on the table, and then three models would be sent backward in time. Then the cycle would continue.

If a time traveler from the year 2200 journeys 50 years into the long ago and then decides to stay there, then the past becomes the "present." As that "present" from the moment of his arrival in 2150 progresses forwards 100 years, day by day year by year, to the future date and time when the time traveler left on his journey in 2200, he will leave again for the past which will, for him and all those living at that time in the past, will be the "present" when he arrives. And as 50 years from his day of arrival go by, day by day, year by year, he will leave again for the past, and then again and again and again such that an infinite number of time travelers might arrive simultaneously in "a" or "the" past or separately in

Quantum Physics of Time Travel

multiple altered pasts which are infinite in number.

The solution to this paradox is quantum physics and the "Many worlds" interpretation. There is more than one past, more than one future, and each time the time traveler repeats the cycle he is not sent to the same past, but to an alternate past, another world, which exists in parallel with innumerable other worlds, each with their own past, present, and future.

17: The Mirror of Time: Red Shifts, Blue Shifts and Duality

Red Shifts, Blue Shifts

The Time Traveler heading into the past is, in some respects, a mirror image moving in reverse. However, an observer on Earth moving forward in time, will mistakenly perceive the time traveler as moving forward in time, but with the tail end of the time machine leading the way--but this is an illusion, a reversal of what appears to be forward vs backward motion (like watching a movie in reverse of a rocket soaring through space). However, in contrast to a movie, from the perspective of an outside observer, there will be two major clues that the time traveler is heading into the past; i.e. the red shifts and blue shifts associated with the movement of the time machine.

If a fire engine is racing down the street the sound and frequency of its siren is higher in pitch as the fire engine approaches the observer and lower after it passes by. However, the emitted frequency never changes (it sounds the same for those riding on the firetruck).

When the source of the sound waves is moving toward the observer, each successive wave crest is emitted from a position closer to the observer than the previous wave, meaning it takes less time to reach the observer. The sound waves contract. Conversely, as the fire truck drives away from the observer, each wave is emitted from a location increasing further from the observer than the previous wave. It takes each successive wave longer to reach the observer, thereby reducing the frequency and stretching them out. When the distance between successive wave fronts is increased, so the waves "spread out" there is a change in the frequency of sound; and the same thing happens with light. This is known as the Doppler Effect. Red shifts and blue shifts are an example of the Doppler effect.

Light from a star as it approaches the event horizon of a black hole, becomes lower and lower in frequency and increasingly red as it accelerates and recedes further and further away from an observer and closer to the hole (Giess, et al., 2010; Melia, 2003a,b; Merloni & Heinz, 2008). Light gets redder as frequency decreases and the waves become further apart relative to an observer. "Redder" results from an increase in the space between each wavelength – equivalent to a lower frequency and lower photon energy. If the source of light is moving away from an observer, then redshift ($z > 0$) occurs. If the source moves towards the observer, the waves become closer together and a blueshift ($z < 0$) occurs. This is true for all electromagnetic waves and is explained by the Doppler effect.

Quantum Physics of Time Travel

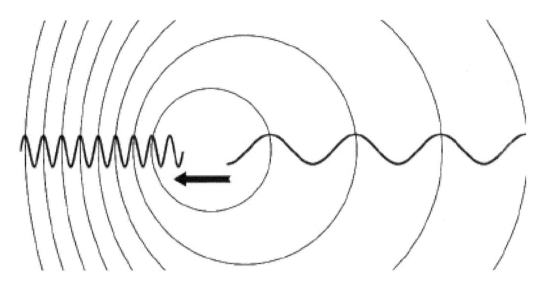

As first theorized by Einstein in 1905, the energy of a photon is inversely proportional to its wavelength. Longer wave lengths have little energy whereas short-wave lengths (such as X-rays) have a lot of energy. For example, if the photon loses 50% of its energy, it's wave length may increase and become up to 10 times larger and display an extreme red shift; that is, the space/distance between the crests of each wave become further and further apart. By contrast, as light approaches an observer, the frequency is increased, the crests of the waves become closer together, and the light displays a blue shift; becoming increasingly blue as the waves crowd together.

As the Time Traveler accelerates toward light speed, light waves from stars in front of the time machine would begin to crowd together, making the star light appear blue. His field of view would also become increasingly narrow and shrink to a tunnel-shaped "window" filled with blue star light which has bunched together. If he looked behind him, the stars would appear to be spreading out and turning red, as if they are accelerating away into the distance.

Upon accelerating toward light speed, an observer on the ground would see the approaching time machine as giving off a blue light, whereas when it shoots passed the observer, it would give off a red light which becomes increasingly red and faint until disappearing. Upon reaching light speed it would give off neither red nor blue as time would seem to stand still.

As the time machine accelerates beyond light speed and heads from the future into the past, then from the perspective of an observer, its red and blue shifts would appear to be reversed--but again, this is an illusion due to time reversal, like watching it go by in a mirror. This is because the observer will believe the time machine is moving forward in time, in lock-step with the observer, when it is coming from the future and heading back in time. However, with each success "now" and "present moment" the time machine is still visible, so the observer thinks it is going forward, when the observer is merely catching up with where the time machine has already been.

The key to the resolving the illusion is the red and blue shifts. It will appear to be red shifted when it appears to approach the observer and blue shifted when traveling further away--but this is an illusion, a consequence of perceiving it as moving forward in time when it traveling backward in time.

For example, if the observer were gazing into the heavens and noticed the time machine passing by overhead at 10 AM that morning, although the time machine is going into the past, the observer would also see it at 10:01 AM and at 10:02 AM, 10:03 AM, 10:04 AM 10:05 AM and so on as the time machine travels from the future toward the present and the past. The Time Machine would appear (to that observer) as if it is continually present but it will appear as if going tail-forward as the observer moves forward in time. The time machine from the future that is heading into the past, would appear to be going back to the future from the perspective of the observer whose sense of time leads from the present to the future and not from the present to the past. The observer would be observing a "movie" going in reverse, but would not know it.

The Time Machine going from the future into the past would be emitting light with the same unchanging frequency. However, the received frequency would be reversed from the perspective of an observer on the ground as it seemingly passes by in one direction, when in fact it is going the other direction. That is, from the observer's perspective, the Time Machine is not headed into the past, but is sharing the eternal now and like the observer appears to be heading into the future (when it is going the opposite direction from the future to the past). From the observer's perspective, the red and blue shifts are reversed. Instead of appearing blue as it approaches from the future it is giving off a red shift. Instead of appearing red as it speeds away it is giving off a blue shift; and this is because it is heading in a direction opposite to that of the observer in space-time. Since it is heading into the past, the Time Machine is red shifted as it appears from the future and becomes blue shifted as it continues into the past.

On the other hand, since the Time Machine traveling into the past has accelerated beyond the speed of light, it may not be seen by an observer until the light carrying the image of the Time Machine catches up with the observer; and this is because the light emitted or reflected by the Time Machine is following behind the Time Machine which is traveling faster than light speed. That is, the time machine itself, may be invisible as its reflected light is trailing behind it. Therefore, what the observer is looking at is not the real time machine, for the time machine is well in front of its light-image. Instead, the observer sees a light-image of the Time Machine which is speeding by at the velocity of light, whereas the Time Machine is going faster than the speed of light--so the light is always falling further and further behind. Consider, for example, a flash of lighting which is followed a few seconds later by the sound of thunder. Light travels faster than sound. However, faster than light speed is faster than light, so the

Time Machine arrives first from the future, continues into the past, and is followed by its light-image.

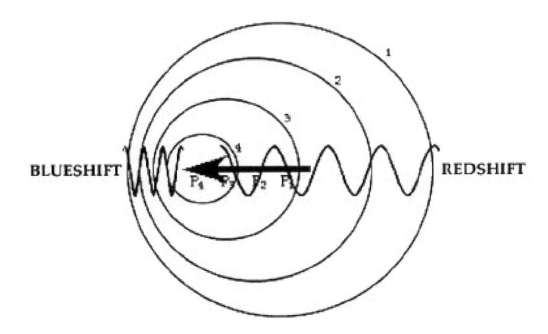

Moreover, when the Time Machine is observed it may appear to be going in two opposite directions at once. This is because the Time Machine had to first journey to the future to travel into the past. From the observer's perspective, the Time Machine may be continually splitting into blue shifts and red shifts as it hurtles into the future and into the past in parallel. The observer may also see multiple Time Machines, the one that has traveled from the past into the future and the one that has journeyed from the future into the past at faster than light speed, followed by a cascade of blue and red shifts. In fact, as the observer moves forward in time, he may see a kaleidoscope of Time Machines, like an accordion of Time Machines, one after the other, the Time Machine arriving from the future at the present moment, the Time Machine at a more distant location heading into the past, and the Time Machine in the present-moment which is headed toward the future; Time Machines both coming and going.

Singularity, becomes duality, which may become a multiplicity. On the other hand, the Time Machine going into the past will probably be microscopic and consisting of negative energy and mass, whereas the Time Machine headed toward the future will have also undergone length contraction, the observer may only notice some strange flashes of blue and red light streaking across the heavens and an area of darkness corresponding to the Time Machine that is traveling faster than light.

Relativity, Space Time...

18: Into the Past: Duality, Anti-Matter and Conservation of Energy

Although time travel to the future poses its own paradoxes, duality is not necessarily one of them until the moment the Time Traveler exceeds light speed. Once the time machine breaks the cosmic speed limit and hurtles into the past, duality is just one of the many paradoxes and conundrums which may result. The Time Traveler will be both coming and going and like other superluminal objects his dual presence in the past and the future could violate the laws of conservation and causality: the future can effect affect the present and the past and there is a potential increase in energy and mass in the cosmos.

The counterpart to the Time Traveler who voyages into the past will have already lived in at least part of that past and will still be in that past. Thus there will be two copies of the Time Traveler, creating duality, and additional mass and energy.

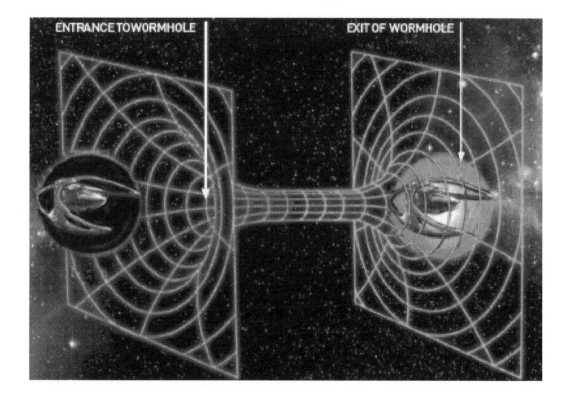

Quantum Physics of Time Travel

For example, upon accelerating from 99.999999999999999% to 100.00000000000001% light speed the time traveler may come face to face with himself (seeing himself perhaps both coming and going), or he could turn his head and gaze in the direction from which he journeyed and see himself following right behind, whereas he would also see himself turning his head. The Time Traveler headed for the future will see himself going the opposite direction in space-time, and the Time Traveler headed for the past will also see himself going in the direction which he already travelled.

At some point, the backward in Time Traveler will speed past the point where his counterpart had not yet left for the future; and the voyage will continue as his counterpart gets younger, is back in college, then high school, then a child, infant, at birth, neonate, fetus, embryo, sperm and egg... molecules...and then there is good ol' mom and dad as he become a twinkle in his father's eye.

A time traveler exceeding the speed of light will not only retrace his own life-line back in time but catch up with the beams of light from the moon, the sun, and Alpha Centauri at the moment they left for Earth, and then back, back, further in time, to before there was a moon or an Earth, or the sun, this solar system. And then, perhaps, if time is a circle, he will eventually circle around from the past and head back for the future. Of course, all this depends on at what superluminal speed he is traveling.

Traveling into the past can be slow going. For example, at 125% light speed, it would take 34 days in the time machine just to go 35 days (one extra day) into the past. If he increased his velocity to 150% light speed, it would take 40 days to travel 2 additional days into the past. Upon accelerating to 180% light speed, then every day in the time machine would take the time traveler 2 days into the past. At 199.9% light speed, the time traveler could leap back 6 years in time for every day in the time machine. Accelerating to 199.999% light speed he could journey 2,000 years back in time in a single day. However, at 200% light speed, time would stop. The Time Traveler would again be upon the event horizon of eternal now which separates the past from the future. At 200.1% light speed, the Time Traveler will be heading back for the future.

On the other hand, he may be able to slow to below light speed and head back to the "present." But in so doing he creates yet another paradox. Upon traveling to the past and then back to the future the time traveler alters every local moment of the quantum continuum and creates a duality which becomes a multiplicity as well as an alternate reality as predicted by the "Many Worlds" interpretation of quantum mechanics (Dewitt 1971; Everett 1956, 1957).

Duality and the Laws of Conservation of Energy and Mass

Duality (and then multiplicity) does not occur upon traveling to the future, but only upon exceeding light speed. The time traveler will be both coming and going once she reaches 100.00000001 light speed. She will be heading into the future as she travels at near light speed toward the future, and she will be heading from the future into the past at above light speed. She would be heading back down the same path as she climbs up that same path. She would be continually catching up with herself in parallel--like two movies, one being run in reverse beginning at the end of the movie, the other being run from the beginning of the movie to the end, on two adjacent movie screens, simultaneously in parallel.

Time Traveler "A" at 99.99999% c and on his way to the future would see himself ("B") appearing from the future and then disappearing into the past. Time Traveler "B" at 100.00001% would see himself ("A") appearing from the past and disappearing into the future. However this cycle would continue in a backward direction with both existing in parallel sharing the same eternal "now" but heading different directions in time, until Time Traveler "A" disappears back into his mother's womb. Until the moment of conception and impregnation, "A" and "B" would exist simultaneously, in parallel, but in different dimensions of time.

Because the extra-energy (or mass) of time traveler "B" who streams from the future into past in parallel with herself ("A") traveling toward the future, creates a duality; which would be a possible violation of the laws of the conservation of energy and mass, until one half of this duality disappeared. This is because the extra-energy of the time traveler heading toward the future would have to be compensated by a corresponding decrease in the energy-mass of the same time traveler heading from the future to the past. That is, when the time traveler begins heading into the past, he will bring with him all his energy and mass as well as that of the time machine. Although mass can become energy and energy mass there is no way to create additional positive mass and positive energy out of nothing. The conservation of energy is constrained by the first principle of relativity and leads to the famous equation $E = mc^2$.

When the time machine appears in the past it must have its own energy or borrow it. If it suddenly appears in the past, then the universe will experience an increase in energy and an increase in mass: the energy and mass of the Time Traveler and his machine. Correspondingly there must be a decrease in energy and mass otherwise the universe will be in a state of mass/energy disequilibrium and in violation of the laws of conservation (Feynman, 1965, 2011; Pollack & Stump 2001; Slater & Frank 2011).

The Conservation of Mass

A Time Traveler going into the past introduces energy and mass which has no source because the source (the time traveler who has not gone into the past) still exists. The energy/mass of Time Traveler "B" can't be borrowed from Time Traveler "A" unless "A" ceases to exist.

The law of conservation of mass decrees that for any system closed to all transfers of matter and energy, the mass of the system must remain constant over time (Feynman 2011) Mass cannot be added unless it comes from energy which is also conserved. Mass cannot be added or removed and is "conserved" over time; and "time" would include the past, present, and future. The law also implies that mass can neither be created nor destroyed, although it may be rearranged in space or transformed into particles or energy.

Likewise, in physics, the law of conservation of energy decrees that the total energy of an isolated system is conserved and cannot change over time. The mass-energy equivalence theorem and the first law of thermodynamics states that mass conservation is equivalent to total energy conservation, and energy can be neither created nor destroyed. Energy however, can change form and be converted into mass and mass into energy, e.g. kinetic energy, potential energy, and electromagnetic radiant energy (Feynman, 1965, 2011; Pollack & Stump 2001; Slater & Frank 2011). All forms of energy contribute to the total mass and total energy. This balance is known as continuous symmetry and this conservation holds over time, including past and future time.

An electron can borrow energy but must pay it back (Feynman 2011). However, the larger the loan the quicker it must be paid back. For the duration of Planck time, i.e. a ten-million-trillion-trillion-trillionth of a second so much energy can be borrowed that it can warp and blow a hole in space-time thereby releasing that borrowed energy back into the quantum continuum. All that is left is negative mass and negative energy.

As detailed in earlier chapters, upon entering a worm hole or black hole, positive energy would likely be stripped away from the time traveler by the negative energy which prevails within the vacuum of these holes. Likewise, upon attaining light speed, and due to length contraction, rather than gaining infinite mass, the time traveler would shrink to a size smaller than a Planck length at which point due to the increase in gravity and because so much energy would be released a hole would be blown through space time. However, that energy would come from the time traveler, leaving him with zero or less than zero energy and mass; a consequence of length contraction continuing in a negative direction.

Therefore, symmetry is actually maintained, for the Time Traveler is a duality and the positive mass and positive energy is matched by the creation of negative mass and energy upon exceeding light speed; the perfect counterbalance to positive energy and mass. If a Time Traveler passes through a hole in space time he may lose energy and have a negative charge when he emerges such that he consists of negative energy and negative mass, and this means there is no excess mass and energy because the duality has resulted in negative mass and energy.

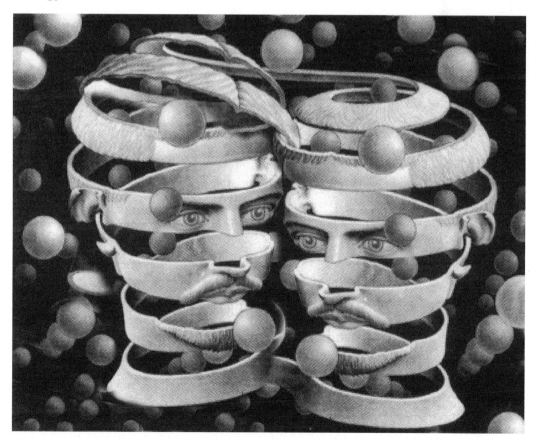

It is precisely because the backwards in time-traveler consists of negative energy and negative mass whereas his duality in the present or future consists of positive energy and mass, that there is no violation of conservation laws.

Unfortunately, the Time Traveler headed toward the past may be in danger of blowing up and disintegrating if he encounters himself; a concept related to "destructive interference" but which is also what would occur if a particle and ant-particle came into contact: mutual annihilation. However, this destructive act would actually release more energy into past-space-time than had existed, again creating an imbalance.

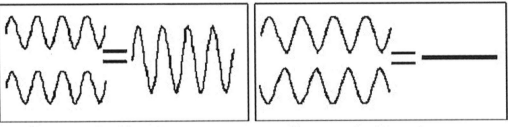

Constructive Interference   Destructive Interference

The Anti-Matter Time Traveler

The Time Traveler headed toward the past from the future, if consisting of negative energy and negative mass, and being the counterpart to the positive energy/mass Time Traveler headed from the past to the future, could be referred to as the anti-Time Traveler. The duality and any resulting asymmetry and violation of conservation laws becomes a problem of matter vs anti-matter.

Asymmetry and the dominance of matter over anti-matter is a fundamental characteristics of the universe (Mazure & Le Brun 2012). There is more matter than anti-matter. The observable, Hubble-Length universe seems to have a nonzero positive baryon number, which is why matter exists. Basically, in physics, a baryon number is a quantum number equal to the difference between the number of baryons and the number of anti-baryons in a system of subatomic particles (Houellebecq 2001). Although Big Bang creationist theory dictates the matter and anti-matter should have been created in equal number and the sum total should be zero (since matter and anti-matter annihilate each other), there is more matter than anti-matter (Mazure & Le Brun 2012).

On the other hand, there may be more anti-matter in the past, thereby creating symmetry with the matter which exists in the future-present.

Anti-matter is composed of anti-particles and has the same mass as ordinary matter but has the opposite charge. Anti-matter has a negative charge and negative mass.

Collisions between particles and anti-particles lead to the annihilation of both, giving rise to the release of varying proportions of high-energy photons (gamma rays), neutrinos, and lower-mass particle–antiparticle pairs in accordance with the mass-energy equivalence equation, $E=mc^2$ (Houellebecq 2001).

Relativity, Space Time...

Antimatter in the form of individual anti-particles, is commonly produced by particle accelerators (Mazure & Le Brun 2012). Particles with negative mass and negative energy are produced in Casimir vacuums (Casimir, 1948; Bressi et al., 2002; Jaffe 2005; Lambrecht 2002) which in turn may characterize the interior of worm holes, super massive black holes, and perhaps even holes smaller than a Plank Length (Everett & Roman 2012; Joseph 2010b,c).

In fact anti-particles are created everywhere in the universe where high-energy particle collisions take place. High-energy cosmic rays impacting Earth's atmosphere produce minute quantities of antiparticles which are immediately annihilated if they come into contact with nearby matter (Mazure & Le Brun 2012). Anti-matter particles are also produced at the center of the Milky Way galaxy, ejected by the supermassive black hole as negatives after positive mass/energy objects fall inside (Mazure & Le Brun 2012). If anti-matter particles are from the future, is unknown.

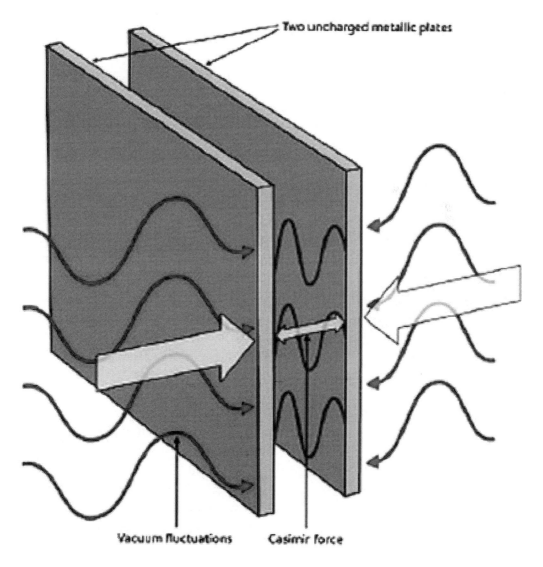

Quantum Physics of Time Travel

Every known kind of particle has a corresponding anti-particle. Therefore, the transformation of a Time Traveler into negative mass/energy would not violate the laws of conservation and would not add significantly to the asymmetry which characterizes space-time as we know it. The danger for Time Traveler "A" and his anti-Time Traveler "B" counterpart, is that they may come into contact. On the other hand, this may be impossible.

If the backwards in time traveler consists of negative mass and negative energy while voyaging at superluminal speeds, he would like be out of phase with the positively charged half of his duality who is traveling at slower than light speeds. In consequence, even if they collide head on they would not be subject to "destructive interference" but instead they may merely pass through each other; like encountering a ghost.

More likely, however, is that they would be unable to make contact, with the negatively charged time traveler being repulsed by the positively charged time traveler who in turn is attracted to the negative charge; the positive energies providing propulsion, the negative energies repulsion. Hence, the positive charged time traveler and the negatively charged time traveler, cancel each other out and there is no violation of the laws of conservancy; although the laws of cause and effect would be directly challenged.

The Laws Of Conservation Are Not Violated By Traveling Into The Past

In quantum mechanics, matter and anti-matter, or positively or negatively charged particles can achieve superluminal values, and journey from the future to the past, such that both exist simultaneously, in parallel, both coming or going. In quantum mechanics, the energy of a quantum system acts on the wave function of the system (Everett 1956, 1957; Neumann 1937). As dictated by the "uncertainty principle" energy and mass can be time-independent (Heisenberg 1927, 1958). This is illustrated by evidence of entanglement where effects may occur simultaneously with causes, and take place at faster than light speeds (Francis 2012; Juan et al. 2013; Lee et al. 2011; Matson 2012; Plenio 2007). As indicated by entanglement, the future may effect or take place before the past-present. Thus, the relationship of cause and effect and energy or mass over time is uncertain and can be described only by probabilities (Born et al 1925, Heisenberg 1925, 1927). Nevertheless, energy at each fixed point in time can be precisely measured in principle. Thus the conservation of energy is a well defined concept even in quantum mechanics.

There is, however, the issue of "closed" vs "open" systems. The laws governing the conservation of energy are enforced only for isolated systems; those completely isolated from all exchanges with the environment (Pollack & Stump, 2001; Slater & Frank, 2011). Mass is not generally conserved in open systems when various forms of energy are allowed into, or out of, the system. In fact, even in a closed system, the principle of matter conservation is not strictly enforced except in the classical sense, and not in respect to special relativity and quantum mechanics.

For example, as demonstrated with the Casimir Force (Casimir, 1948; Bressi et al., 2002; Jaffe 2005; Lambrecht 2002), negative energy and negative mass (along with other virtual particles) may appear in an enclosed vacuum. Their source is unknown, though they may be considered the negative equivalents (or mirror images) of positive energy/mass, such that their sudden existence does not violate the laws of conservation.

The conservation of energy and mass is only approximately true if the time traveler speeding into the long ago consists of negative mass, negative energy and is less than a Plank length in size and/or traveling at superluminal speeds.

Therefore conservation laws are retained and there would be balance and equilibrium as less than nothing was added to the dimension of space-time known as the past by a Time Traveler voyaging into the long ago. There is equivalence because Time Traveler "A" who is heading to the future consists of positive mass and positive energy whereas the reverse is true of "B." However, because he consists of negative energy and negative mass the time traveler heading backwards in time might instead appear as a ghost, or remain smaller than a Plank length and be essentially invisible.

Because the time traveler voyaging into the past consists of negative energy and negative mass, he does not increase either mass or energy of the universe as he journeys through it. Likewise, because his positive energy/mass counterpart is traveling into the future and existed in the past, a balance between positive and negative would continue even before the time traveler is born as the molecules, atoms, particles which gave him shape and form continue to exist even as his counterpart continues to journey into the past.

The Negatively Charged Time Traveler May Never Be Able to Land in the Past

Unfortunately, unless the past consists of negative energy and negative mass, the time traveler will never be able to stop and visit the past. His negative mass and negative energy would be repulsive. To have no mass, to be massless, or negative mass, may be to have no gravity and thus to be anti-gravity. Anti-gravity was predicted by Einstein in 1917 when he proposed a repulsive form of gravitation which would balance the attractive force of gravity. However, as to a negative mass and negative energy structure, it would be repulsed by a positively charged Earth. The Time Machine can fall but it could never land.

And if, while traveling into the past he was able to make contact with the positively charged Earth or some other positively charged objects, he may disintegrate into electromagnetic radiation. On the other hand, in a mirror universe, where everything is in reverse, then perhaps negative charges attract.

It is these conundrums, coupled with miniaturization, which may explain why we are not overrun with tourists from the future.

19: Quantum Entanglement And Causality: The Future Effects the Past

The river of time is bent round in a circle and it may have no beginning and no end. Since space-time is curved, warped and littered with vortexes surrounding black holes and effected by the gravity of innumerable stellar objects, the river of time may also be split apart and bent backwards in a circle, with circles within circles, as happens with whirlpools and eddies along river banks. Likewise, the geometry of time may flow differently in various regions of the cosmos and split off into innumerable tributaries of time each with their own unique trajectory and velocity.

Relativity, Space Time...

Because space is "isotropic" there is nothing in the law of physics indicating that a particular direction is preferred; down, up, sideways, backwards, its all the same. Why should space-time, or time, be any different? Since the past, present and future overlap and are relative to observers and differ according to location, gravity, and speed of movement, then as Einstein stated, the distinctions between them are an illusion.

The trajectory of the arrow of time may be from the future toward the present, and then the flow may continue from the present to the past. And like a flowing river, the "present," "past" and "future" are relative to an observer; whereas in fact, the river has no present or past or future, it just flows as its own unity. If time is a circle, then time is also a unity.

The laws of electromagnetism do not make a distinction between past and future (Pollack & Stump, 2001; Slater & Frank, 2011). And yet, although light waves travel in a direction, it is assumed that these waves are traveling from the present into the future, when in fact they are traveling into the past and the future and from the future and from the past relative to different observers on different worlds and even on the same planet.

A light wave from Earth takes 4.2 light years to reach Proxima. However, since it will not be received on Proxima for 4.2 light years it will not arrive until some future date on Proxima and is in the future relative to observers on Proxima, although from the past relative to those on Earth. Likewise, light-images which just left Proxima are from Proxima's past but will not arrive on Earth until some day in their future. The future and the past are relative. Moreover, once light-images from Proxima arrive on Earth, they continue into the past relative to those on Earth, but not relative to those on a planet 4.2 light years in the opposite direction from Earth and 8.4 light years from Proxima, in which case although they are from the past relative to Earth and Proxima they will not be received until some future date for those denizens of that more distant alien world. This conception of time is entirely consistent with Einstein's theories of relativity and Maxwell's equations of electromagnetism.

The past and future exist simultaneously in different and overlapping locations in space. Since space is isotropic, then, theoretically, there are no roadblocks to prevent a time traveler from choosing a location at will and then speeding into the future or the past; just as he may decide to go up-river or down-river.

The past, present and future, however, are like the weather, and differ in distant locations. There is no universal "now" and there are innumerable pasts, presents and futures which increasingly diverge as distance from any particular observer increases. Time is relative and the same can be said of the future and the past which only remains approximately and generally similar relative to observers sharing the same local, or personal, frames of references. Only when frames of reference are shared locally can observers agree on what took place first and last and what is in the past and what is still in the future.

Quantum Physics of Time Travel

Is The Future Determined? Can The Past Be Changed?

Light can travel to the future and from the past relative to the observer's frame of reference. However, light and time are not the same. The speed of light, and time, be it past or future, are not synonymous, though both may be affected by gravity (Carroll 2004; Einstein 1961). Moreover, just as light has a particle-wave duality and can physically interact with various substances, time also can be perceived and therefore must have a wave function if not a particle-wave duality.

Time-space is interactional, and can contract to near nothingness and then continue to contract in a negative direction such that the time traveler can journey into the past. Therefore, time, and time-space are embedded in the quantum continuum and can effect as well as be effected by other particle-waves even at great distances; a concept referred to as "entanglement." Time and space-time are entangled.

As demonstrated in quantum physics, the act of observation, measurement, and registration of an event, can effect that event, causing a collapse of a the wave function (Dirac 1966a,b; Heisenberg 1955), thereby registering form, length, shape which emerges like a blemish on the face of the quantum continuum. Likewise, a Time Traveler or particle/object speeding toward and then faster than light and from the future into the past will affect the quantum continuum. By traveling into the future or the past, the Time Traveler will interact with and alter every local moment within the quantum continuum and thus the future or the past.

If the past or the future are not altered, this means that these dimensions of time are hardwired as part of the quantum continuum, that these events were already woven into the fabric of time and had always happened and always will happen and cannot be altered because they already happened, albeit in different distant locations of space-time which are linked as a unity within the quantum continuum.

If the future/past are not altered by voyaging through time then this is because the Time Traveler had already journeyed into the past and future before he journeyed into it.

As detailed in previous chapters, by accelerating toward light speed, the distance between the present and the future contracts such that a time traveler can arrive at the future more quickly than those left behind. However, what this implies is that the future already exists; a concept which is intrinsic to space-time relativity. The future and the past exist in various overlapping locations in space-time which are in motion.

Therefore, just as the end of a movie already exists as one begins watching the movie, then perhaps the same may be said of the river of time as related to the future and the past. If this premise is correct, then just as one can't alter the ending or the beginning of the movie, one can't alter the future or the past because it is hard-wired into the quantum continuum of space-time. If true, then the Time Traveler would be unable to go back in time to kill Hitler, prevent the assassination of the Kennedy brothers, or stop the 2001 terrorist attack in New York known as "9-11."

The Laws of Consistency And Uncertainty

The idea that it is impossible to change the past has been referred to by Ignor Novikov and Kip Thorne (Friedman et al. 1990) as "the principle of self-consistency." The past remains consistent because it is hardwired and can't be changed. If the time traveler goes into the past, then he always traveled to the past, even before he built his time machine and journeyed to the past. Therefore, whatever he does in the past, is already part of the past record and woven into the fabric of time. He did it before he did it and will always do it exactly the same.

If the future already exists, then can the same principles of "consistency" apply? Wouldn't this violate the laws of cause and effect? Yes, and no. Newton proposed that if we knew the velocity, position and mass of every particle in the universe we could make accurate predictions about the future until the end of time. By contrast, the Uncertainty Principle holds that we cannot know velocity, position and mass of any particle simultaneously. Moreover, uncertainties propagate into the future; that is small uncertainties becomes larger as distance from the present increases.

Systems are chaotic, like the weather, and small changes can become bigger changes over time, at least locally (Kellert 1993). Then over time and distance their effects lose power and dissipate like a cloud of smoke.

Therefore, if the future already exists, it is not hard wired but subject to change and may be in continual flux, becoming only predicable the closer it comes to the present; like the weather. Thus the present can effect the future but only with a certain degree of probability with the power of predictability decreasing as to more distant futures. The power to predict cause and effect are limited. Thus, we have instead, probabilities.

If chaos theory (Kellert 1993) and the Uncertainty Principle (Heinsenberg 1927) are applied to the past, then perhaps the past is also in flux and subject to change. Therefore, the Time Traveler may be able to go back in time and kill Hitler or prevent the murder of the Kennedy brothers. However, if successful, then, according to the "Many Worlds" interpretation of quantum physics (DeWitt 1971; Everett 1956, 1957), the actions of the Time Traveler did not change "the" past per se, but rather his actions increased the probability that another probable past which always existed, would become his past.

There is no "The" past. There is no "universal" now. Time is relative (Einstein 1955, 1961). The past, present, future occupy overlapping as well as distant locations which are in motion and which interact with their surroundings, creating ripples in the river of time. However, just as rocks tossed at various distances into the smooth surface of a crystal lake create waves which may intersect, time is entangled with the quantum continuum and all of space-time no matter how distant; and that includes the future and the past. In consequence, if correct, then a future "cause" can "effect" the present, and the future and the present can effect the past because all are entangled.

Quantum Entanglement

To get to the past, the Time Traveler must accelerate beyond light speed into the future, and then, upon achieving a velocity above the speed of light, the destination becomes the past. Time is a circle and the future leads to the present, and the past.

It is well established that objects respond to and can influence and affect distant objects at speeds faster than light. This "spooky action at a distance" has been attributed to "fields," "mediator particles," gravity, and "quantum entanglement" (Bokulich & Jaeger, 2010; Juan et al. 2013; Sonner 2013).

For example, it is believed that an electric "field" may mediate "electrostatic" interactions between electromagnetic charges and currents separated by great distances across space. However, these changes can take place at faster than light speeds. Charged particles, for example, produce an electric field around them which creates a "force" that effects other charges even at a distance. Maxwell's theories and equations incorporate these electrostatic physical "fields" to account for all electromagnetic interactions including action at a distance.

Since mass can become energy and energy mass, the "field" is therefore a physical entity that contains energy and has momentum which can be transmitted across space. Therefore, "action at a distance" may be both distant and local, a consequence of the interactions of these charges within the force field they create. However, the problem is: the effects can be simultaneous, even at great distances, and occur faster than the speed of light (Plenio 2007; Juan et al. 2013; Francis 2012; Schrödinger & Dirac 1936), effecting electrons, photons, atoms, molecules and even diamonds (Lee et al. 2011; Matson 2012; Olaf et al. 2003; Schrödinger & Born 1935). The effect, therefore, may precede the cause since it takes place faster than light.

Consider entangled states: two particles which are far apart have "spin" and they may spin up or down. However, although they are far apart, an observer who measures and verifies the spin of particle A will at the same time effect the spin of particle B, as verified by a second observer. Measuring particle A, effects particle B and changes its spin. Likewise observing the spin of B determines the spin of A. There is no temporal order as the spin of one effects the spin of the other simultaneously. Even distant objects are entangled and have a symmetrical relationship and a constant conjunction (Bokulich & Jaeger, 2010; Plenio 2007; Sonner 2013). If considered as a unity with no separations in time and space, then to effect one point in time-space is to effect all points which are entangled;

and one of those entangled connections is consciousness (Joseph 2010a). And this gives rise to the uncertainty principle (Heisenberg, 1927). Correlation is not causation and it can't always be said with certainty which is the cause and which is the effect and this is because the cosmos is entangled.

World Lines, Causality, and Entanglement

Time, too, can be entangled. Future, past, present, are relative and overlap, and what is the future in one galaxy can be the past in another; all are entangled in the fabric of space-time and the quantum continuum.

If time is conceived as a spatial gestalt, an interconnected continuity of length, width, height, and extent but without temporal order, then what takes place in one location of space-time can effect what takes place in another, even if the distance is measured in miles, minutes or hours.

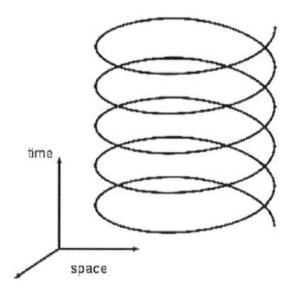

For example, in the Great Basin, White Mountains of California, there are "bristlecone pine" trees over 5000 years old, and which stand over 50 feet high. However, if the tree is measured in space-like intervals and not time-like intervals these same trees could be viewed as having a length of 5000 years. That is, if its "world line" was visualized as a thick strand of rope moving through space, that rope would begin with the seed and extend to the top of the tree. Likewise, if the orbit of Earth was viewed as a strand of rope, that rope would circle around the sun; for in fact, the movement of Earth is a continuity and is not separated into intervals which take place one after another like the ticking of a clock.

The "world line" of the tree, for example, encompasses it's entire history and although the tips of some branches and roots may have only recently grown, they are connected with the entire tree from the roots to the crown, and thus to the youngest and oldest parts of the tree. And what takes place in the roots can effect the twigs, branches, and crown of the tree, and the condition of the crown can effect the roots, branches and twigs. If viewed from a space-like intervals, the seeds of the tree and the 50 foot tree becomes an interconnected continuity.

Consider the evolution of life on this planet, beginning with the DNA of the first living creatures to take root on Earth, e.g. archae, bacteria, and viruses; most or all of whom are believed to have engaged in horizontal gene transfer and other biological activities which eventually led to the step-wise progression leading from single cells to woman and man. .

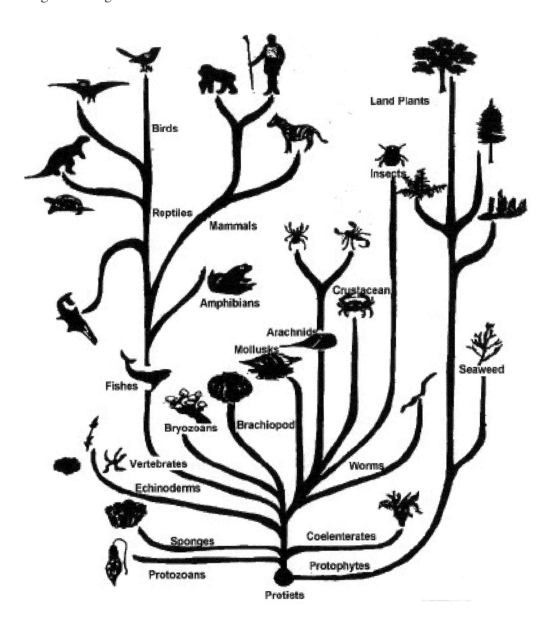

Relativity, Space Time...

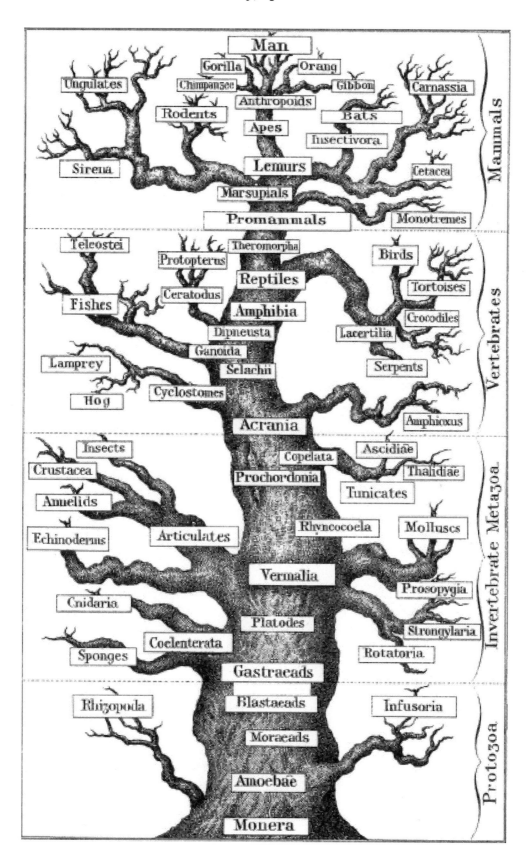

Quantum Physics of Time Travel

Most scientists agree that all living creatures evolved from these first Earthlings, and are linked genetically. Therefore, from the perspective of DNA there is a continuity which extends backwards in time to at least 4.2 billion years or longer. Although genes have been deleted and new genes fashioned, there reamins a core genetic heritage. Thus the tree of life could also be considered as an unbroken string which could be described as 4.2 billion years in length.

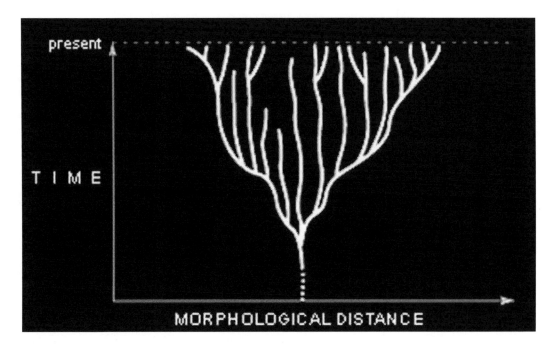

If time is considered from the perspective of space-like intervals and not time-like intervals, then causality can be forward, backward, or simultaneous (Bonor & Steadman, 2005; Buser et al. 2013; Carroll 2004; Gödel 1995). The future and the past become entangled as a continuity in space-time.

World Lines of Time, Causality, Entanglement

If the "world line" of time is viewed as a rope that extends into the future from the past, then it would be possible for a time traveler to loop backwards in time, and intersect his own world-line and come face-to-face with himself, and then travel further back in time and change and effect events which took place in the past.

If events are separated by a time-like interval there may be a causal relationship between them, e.g., what happens first, second, third, or A causes B which causes C. However, given a reference frame in which two events are separated by a space-like interval they may be entangled and occur in different regions of space and space-time but without any causal relationship although they occur together and simultaneously as if linked as one (Bokulich & Jaeger,

2010; Lee et al. 2011; Matson 2012; Olaf et al. 2003). Thus two events may occur in different locations of space-time, but not necessarily in each other's future or past.

For example, the trajectories of elementary (point-like) particles through space and time are a continuum of events called the "world line" of the particle (Caroll 2004; Gödel 1995). Extended or composite objects (consisting of many elementary particles) are a union of many world lines twisted together by virtue of their interactions through space-time. Therefore, the past, present and future are linked in space and in time and can even intersect themselves.

A fetus which becomes a neonate, then an infant, child, adolescent, adult, can be viewed in terms of a continuum through space, like a string which stretches from conception to death. Or the growth of the fetus can be viewed in time-like intervals: 2 months post-conception, then 9 months of age, 3 years of age, 15 years of age, 21 years of age, 70 years of age. However, if viewed as a continuous world-line, as a spatial gestalt, instead of measuring the person in time-like intervals and saying the person is 70 years old, they could be characterized as 70 years in length, and we would see the person as an embryo-fetus-neonate-child-adolescent-adult, simultaneously; and this could be referred to as that person's "world line." When considered as spatial, and not as time-like segments, the 3-year old child is not a separate entity from the 70 year old man, no more than the root of a tree is separate from the branches towering overhead.

However, if someone were to cut of the top of the tree and all its branches, and if other trees grew above it so it no longer received any sun, then the entire tree, including its roots, may die. Considered as a gestalt in space, the death is a continuity. If considered in time like intervals, it could be said that the future (the top of the tree) effected the past (the base of the tree).

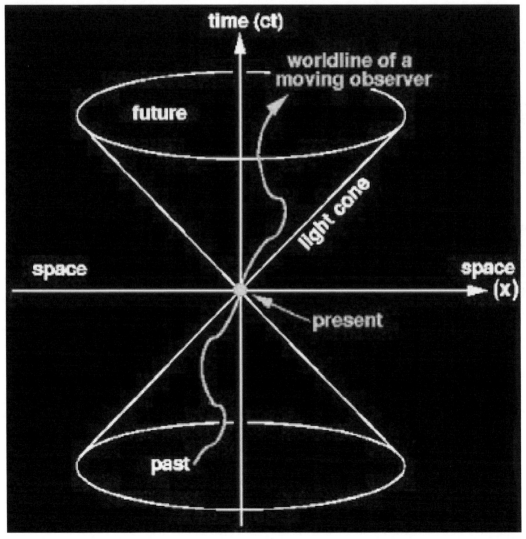

If the time traveler and his time machine were viewed not as a series of objects located in separated points at different times, but as a continuity, like a strand of string, then that string would lead from the past to the present to the future and from the future to the past thereby deforming all aspects of space-time locally and effecting space-time at a distance because past/present/future are entangled and separated by space, not time (Bokulich & Jaeger, 2010). Hence, if one were to travel from 2100 to 2200 and from 2200 to 1900, then every moment, every second, every minute, hour, day, week, month, year, leading up to 2200 and then back to 1900 would be effected and altered as well as connected by the "string", the "world line" of the time traveler. Moreover, the effects can ripple to other locations in space-time such that events before 1900 and well after 2200 are effected, like ripples cascading across the surface of a clear blue lake, due to entanglement.

As based on quantum mechanics, it could be said that in space-time all things overlaps and coincide and exist side by side, including time. If time is considered as spatial and different aspects of time are separated only by distance, then the future, past, and present coexists simultaneously, albeit in different locations in space which overlap and are continually in flux. Thus future, past, present, and time, can be viewed as the union of all events in the same way that a line is the union of all of its points. What effects the future, can therefore effect the past as a continuity which is linked and not broken up and into isolated fragments which are separated by artificial time-like intervals. Likewise, when considered as phenomenon taking place in space, then the past can also effect the future as they are entangled as a unity.

Causality, as a form of entanglement, a singularity which responds as a unity, is preserved when considered in space-like intervals. Causality, however, may be violated when viewed in time-like intervals, for so called effects can occur faster than the speed of light (Lee et al. 2011; Matson 2012; Olaf et al. 2003). As demonstrated with entanglement (Francis 2012; Juan et al. 2013; Plenio 2007), the future may effect the past even though past and present are separated by varying distances in space time. Quantum entanglement is a feature of time.

Quantum Entanglement, Gravity and Superluminal Time Travel

Research into quantum entanglement was initiated by a 1935 paper by Albert Einstein, Boris Podolsky, and Nathan Rosen describing the EPR paradox (Einstein et al. 1935) the purpose of which was to discredit and disprove the Copenhagen interpretation of quantum physics. Instead, entanglement was verified experimentally (Bohr, 1958; Schrödinger & Dirac 1936) thereby disproving Einstein's theory that nothing can travel faster than light.

Specifically, Einstein and his colleagues (Einstein et al. 1935) proposed a thought experiment involving position and momentum: Two particles speed off in opposite directions and travel so far distant that it should be impossible for there to be any "classical" interactions between them; as based on the principle of locality and his special theory of relativity. And yet, despite Einstein's claims, when these experiments were carried out, it was found, repeatedly, that the measurement of one particle effected and determined the measurement of the other at superluminal speeds (Francis 2012; Juan et al. 2013; Lee et al. 2011; Matson 2012; Olaf et al. 2003; Plenio 2007); evidence for non-locality and against cause and effect, exactly as predicted by quantum mechanics. The behavior of distant objects are "entangled" as their changing behaviors occur simultaneously as a single event no matter how distant they are from one another (Bokulich, & Jaeger 2010). The same principles may apply to time.

Entanglement has been repeatedly scientifically verified (Francis 2012; Juan et al. 2013; Lee et al. 2011; Matson 2012; Olaf et al. 2003; Plenio 2007); which raises the possibility that something that happens in America can instantly effect something happening in China, or that an event in this galaxy can cause an instantaneous change in another galaxy. Likewise, if time is entangled, then just as the present can effect the future, the future may effect the present, and the past.

Entanglement is directly related to Heisenberg's (1927) Uncertainty Principle and to what Einstein (1955) called "spooky action at a distance." Not only does entanglement and simultaneity violate the upper limit of the speed of light but its a demonstration that information can travel faster than light speed. As detailed in earlier chapters, the only way to break Einstein's cosmic speed limit is to first journey to the future by accelerating toward the velocity of light and it is only upon exceeding light speed that space-time contraction continues in a negative direction, and thus from the future into the past.

In the typical entanglement experiment, an event in one location can have an effect on another structure at faster than light speeds. However, what this could also imply is that the reaction of the distant structure may have occurred first. Since it takes place at superluminal speeds then it must have occurred in the future and then sped into the present.

The implications of entanglement are that effects can take place before their cause and thus the future may occur before the present by sending information into the past from the future at faster than light speeds; contrary to the laws of causality and Einstein special theory of relativity.

Gravity, Time, And Entanglement.

It was precisely because of superluminal effects that Einstein proposed his general theory of relativity and gravity. According to Newton's theory, gravity acts as a force which can effect distant objects. Einstein (1913, 1914, 1915a,b) rejected this view. According to Einstein's (1961) theory of special relativity, nothing can have an instantaneous effect on a distant object since, according to his theorem, nothing can travel faster than the speed of light. Since the information or force acting on distant objects must be conveyed through space then it must be perturbations and alterations in space and the geometry of space-time, and thus the contiguity of distant objects connected by space-time which accounts for instantaneous effects.

As theorized by Einstein (1915a,b, 1961), the presence of matter causes a depression in space-time, warping its geometry. The presence of matter causes space-time to become non-Euclidian. Therefore changes and deformations in space-time geometry can inform objects, at a distance, of these changes. Matter acts on space-time and alterations in space-time act on matter. Therefore, objects approaching or entangled with these curvatures respond instantaneously.

Since time has energy, and as energy can become matter, and as time may have a particle-wave duality, then time should also be able to effect and warp the geometry of space-time. Does time also have gravity?

Entanglement and "spooky action at a distance," however, are not a function of gravity. Rather entanglement is a reflection of the continuity of space-time and the experience of time when measured as a spatial phenomenon of continuity and not as a succession of isolated temporal sequences. If the world-line of time is conceptualized as a string, then no matter where the string is plucked, the entire string will vibrate.

Again, the same can be said of time. Einstein (1955) admitted that the distinction between past, present and future are an illusion. All are linked and overlap in space-time.

Since space-time includes "time," then if space-time consists of energy which can become matter and all of which are effected by gravity, then time is also interactional, which is why it can be experienced and perceived. Time can act on matter and time has energy which can become matter, a particle-wave duality which propagates through space. This implies that time can also effect and warp the space-time continuum which includes multiple futures and multiple pasts which share world lines which can overlap and intersect one another. Time, the future, the past, all are entangled. The river of time is a circle.

20: Light, Wave Functions and the Uncertainty Principle: Changing the Future and the Past

Cause, Effects and The Uncertainty Principle

In 1904, Lorentz introduced a hypothesis that moving bodies contract in their direction of motion by a factor depending on the velocity of the moving object. He also argued that in different schemes of reference there are different apparent times which differ from and replace "real time." He also argued that the velocity of light was the same in all systems of reference.

In 1905 Albert Einstein seized on these ideas and abolished what Lorentz called "real time" and instead embraced "apparent time." In his theories of special relativity, Einstein promoted the thesis that reality and its properties, such as time and motion had no objective "true values" but were "relative" to the observer's point of view (Einstein, 1905a,b,c). Einstein's conceptions of reality, therefore, differed significantly from that of Newton.

Einstein's theories did not replace Newtons. Instead Einstein came up with a new closed system of definitions and axioms represented by mathematical symbols which were are radically different from those of Newton's mechanics. For example, space and time in Newtonian physics are independent, whereas in relativity they are combined and connected by the Lorentz transformations. Moreover, although Newtonian mechanics could be applied to events where velocities are small relative to the velocity of light, Newtonian physics cannot be applied to events which take place near light speeds whereas Einstein's physics can. By contrast, it is at light speed and beyond, and for objects and particles smaller than atoms where Einstein's theory breaks down and this was recognized in the early 1920s (Born et al. 1925; Heisenberg 1925, 1927). The phenomenon of electricity, electromagnetism and atomic science required a new physics.

In 1925 a mathematical formalism called matrix mechanics posed a direct challenge to Newton and Einstein (Born et al. 1925; Heisenberg 1925). The equations of Newton were replaced by equations between matrices representing the position and momentum of electrons which were found to be unpredictable. Broadly considered, atoms consist of empty space at the center of which is a positively charged nucleus and which is orbited by electrons. The positive charge of the atom's nucleus determines the number of surrounding electrons, making the atom electrically neutral. However, it was determined that it was impossible to make precise predictions about the position and momentum of electrons based on Newtonian or Einsteinian physics, and this led to the Copenhagen interpretation (Heisenberg 1925, 1927) which Einstein repeatedly attacked because of all the inherent paradoxes. Matrix mechanics is referred to now as quantum mechanics whereas the "statistical matrix" is known as the "probability function;" all of which are central to quantum theory.

As summed up by Heisenberg (1958) "the probability function represents our deficiency of knowledge... it does not represent a course of events, but a tendency for events to take a certain course or assume certain patters. The probability function also requires that new measurements be made to determine the properties of a system, and to calculate the probable result of the new measurement; i.e. a new probability function."

Quantum physics, as exemplified by the Copenhagen school (Bohr, 1934, 1958, 1963; Heisenberg, 1925, 1927, 1930), like Einsteinian physics, makes assumptions about the nature of reality as related to an observer, the "knower" who is conceptualized as a singularity. As summed up by Heisenberg (1958), "the concepts of Newtonian or Einsteinian physics can be used to describe events in nature." However, because the physical world is relative to being known by a "knower" (the observing consciousness), then the "knower" can influence the nature of the reality which is being observed through the act of measurement and registration at a particular moment in time. Moreover, what is observed or measured at one moment can never include all the properties of the object under observation. In consequence, what is known vs what is not known becomes relatively imprecise (Bohr, 1934, 1958, 1963; Heisenberg, 1925, 1927).

As expressed by the Heisenberg uncertainty principle (Heisenberg, 1927), the more precisely one physical property is known the more unknowable become other properties. The more precisely one property is known, the less precisely the other can be known and this is true at the molecular and atomic levels of reality. Therefore it is impossible to precisely determine, simultaneously, for example, both the position and velocity of an electron (Bohr, 1934, 1958, 1963).

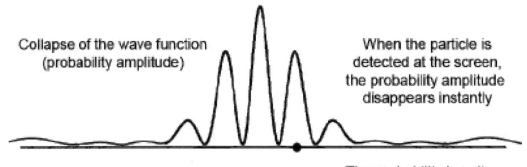

Relativity, Space Time...

Heisenberg's principle of indeterminacy focuses on the relationship of the experimenter to the objects of his scientific scrutiny, and the probability and potentiality, in quantum mechanics, for something to be other than it is. Einstein objected to quantum mechanics and Heisenberg's formulations of potentiality and indeterminacy by proclaiming "god does not play dice."

In Einstein's and Newton's physics, the state of any isolated mechanical system at a given moment of time is given precisely. Numbers specifying the position and momentum of each mass in the system are empirically determined at that moment of time of the measurement. Probability never enters into the equation. Therefore, the position and momentum of objects including subatomic particles are precisely located in space and time as designated by a single pair of numbers, all of which can be determined causally and deterministically. However, quantum physics proved that Einstein and Newton's formulation are not true at the atomic and subatomic level (Bohr, 1934, Born et al. 1925; Heisenberg 1925, 1927).

According to Heisenberg (1925, 1927, 1930), chance and probability enters into the state and the definition of a physical system because the very act of measurement can effect the system. No system is truly in isolation. No system can be viewed from all perspectives in totality simultaneously which would require a god's eye view. Only if the entire universe is included can one apply the qualifying condition of "an isolated system." Simply including the observer, his eye, the measuring apparatus and the object, are not enough to escape uncertainty. Results are always imprecise.

As determined by Niels Bohr (1949), the properties of physical entities exist only as complementary or conjugate pairs. A profound aspect of complementarity is that it not only applies to measurability or knowability of some property of a physical entity, but more importantly it applies to the limitations of that physical entity's very manifestation of the property in the physical world. Physical reality is defined by manifestations of properties which are limited by the interactions and trade-offs between these complementary pairs. For example, the accuracy in measuring the position of an electron requires a complementary loss of accuracy in determining its momentum. Precision in measuring one pair is complimented by a corresponding loss of precision in measuring the other pair (Bohr, 1949, 1958, 1963). The ultimate limitations in precision of property manifestations are quantified by Heisenberg's uncertainty principle and matrix mechanics. Complementarity and Uncertainty dictate that all properties and actions in the physical world are therefore non-deterministic to some degree.

Bohr (1949) holds that objects governed by quantum mechanics, when measured, give results that depend inherently upon the type of measuring device used, and must necessarily be described in classical mechanical terms since the measuring devices functions according to classical mechanics. The measuring device effects the outcome and the interpretation of that outcome as does the observer using that device. "This crucial point...implies the impossibility of any sharp separation between the behaviour of atomic objects and the interaction with the measuring instruments which serve to define the conditions under which the phenomena appear...." (Bohr 1949).

Evidence obtained under a single or under different experimental conditions cannot be reduced to a single picture, "but must be regarded as complementary in the sense that only the totality of the phenomena exhausts the possible information about the objects." In consequence, the results must be viewed in terms of probabilities when applied to the nature of the object under study and its current and future behaviors. Bohr (1949) called this the principle of complementarity, a concept fundamental to quantum mechanics and closely associated with the Uncertainty Principle. "The knowledge of the position of a particle is complementary to the knowledge of its velocity or momentum." If we know the one with high accuracy we cannot know the other with high accuracy (Bohr, 1949, 1958, 1963; Heisenberg, 1927, 1955, 1958).

Central to the Copenhagen principle is the wave function and the probability distribution, i.e. the results of any experiment can only be stated in terms of the probability that the momentum or position of the particles under observation may assume certain values. The probability distribution is a prediction for what may occur in the future, that is, within a predicted range of probabilities. When the experiments are performed many times, and although subsequent observations may differ, they are expected to fall within the predicted probability distribution. This also means that nothing is precisely determined (Bohr, 1949, 1963; Heisenberg, 1927, 1930, 1955).

Hence, whereas Einstein and Newton embrace determinism and causality, Heisenberg (1925, 1927; 1958) argues that although every deterministic system is a causal system, not every causal system is deterministic. Rather, causality is the relationship between different states of the same object at different times whereas what is "deterministic" relates to what may occur, and is better described in terms of probabilities.

Time too, however, is also uncertain. Temporal succession may have no probable connection with what precedes or follows (Heisenberg 1958). In quantum mechanics, one can know the connection only by knowing the future state--thus one must wait for the future to arrive, or look back upon the future state of similar systems in the past. If one knows the properties of an acorn at an earlier time t1 one still cannot deduce the properties of the oak tree at time t2. This may be possible only in isolated systems (Bohr, 1949; Heisenberg 1958). However, unless the entire universe is included in the measurement then the system is not truly isolated.

Quantum mechanics is mechanical but not deterministic and causal relationships are never teleological and not always deterministic. In quantum physics, nature and reality are represented by the quantum state. The electromagnetic field of the quantum state is the fundamental entity, the continuum that constitutes the basic oneness and unity of all things. The physical nature of this state can be "known" by assigning it mathematical properties and probabilities (Bohr, 1958, 1963; Heisenberg, 1927). Therefore, abstractions, i.e., numbers and probabilities become representational of a hypothetical physical state. Because these are abstractions, the physical state is also an abstraction and does not possess the material consistency, continuity, and hard, tangible, physical substance as is assumed by Classical (Newtonian) physics. Instead, reality, the physical world, is a process of observing, measuring, and knowing and is based on probabilities and the wave function (Heisenberg, 1955).

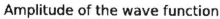

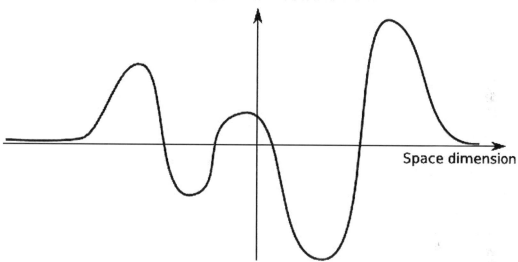

Quantum Wave Function

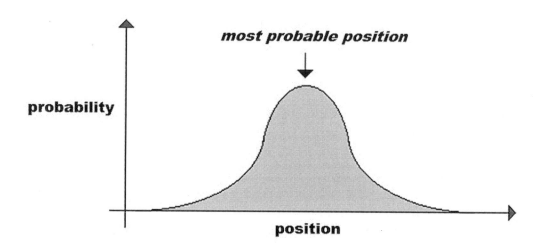

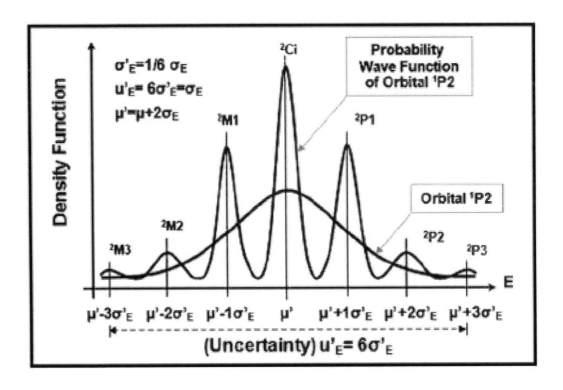

Consider an elementary particle, once its positional value is assigned, knowledge of momentum, trajectory, speed, and so on, is lost and becomes "uncertain." The particle's momentum is left uncertain by an amount inversely proportional to the accuracy of the position's measurement which is determined by values assigned by measurement and the observing consciousness. Therefore, the nature of reality, and the uncertainty principle is directly affected by the observer and the process of observing, measuring, and knowing, all of which are variable thereby making the results probable but not completely certain (Heisenberg, 1955, 1958).

"What one deduces from an observation is a probability function; which is a mathematical expression that combines statements about possibilities or tendencies with statements about our knowledge of facts....The probability function obeys an equation of motion as the co-ordinates did in Newtonian mechanics; its change in the course of time is completely determined by the quantum mechanical equation but does not allow a description in both space and time" (Heisenberg, 1958).

"The probability function does not describe a certain event but a whole ensemble of possible events" whereas *"the transition from the possible to the actual takes place during the act of observation... and the interaction of the object with the measuring device, and thereby with the rest of the world... The discontinuous change in the probability function... takes place with the act of registration, because it is the discontinuous change of our knowledge in the instant of*

registration that changes the probability function." "Since through the observation our knowledge of the system has changed discontinuously, its mathematical representation has also undergone the discontinuous change and we speak of a quantum jump" (Heisenberg, 1958).

Einstein ridiculed these ideas: "Do you really think the moon isn't there if you aren't looking at it?"

As theorized by Einstein (1961), and unlike the Copenhagen model of quantum physics (Bohr 1949; Heisenberg 1927, 1958), space-time is relative to but independent of any observer. Consciousness and the act of measurement is relative but irrelevant having no effect on the passage of time or events. In relativity, each event which occurs at certain moments of time in a given region of space are relative to those observers in different regions of space, such that each observer chooses a convenient metrical coordinate system in which these events are specified by four real numbers. All four dimensions are measured in terms of units of distance, e.g. in two spatial dimensions x and y and a time dimension orthogonal to x and y.

In an Einsteinian universe the observed rate at which time passes for an object depends on accelerated frames of reference and the strength of gravitational fields; all of which can slow or accelerate the passage of time, depending on the object's velocity relative to the observer (Einstein 1961). Specifically, time slows at higher speeds in one reference frame relative to another reference frame. The duration of time can therefore vary according to reference frames.

In relativity consciousness is merely relative. In quantum physics, consciousness and the act of observation and measurement constitute a separate reference frame which can collapse the wave function and register entangled interactions within the environment. Consciousness by the act of observation or measurement takes a static or series of *pictures-in-time* which then becomes discontinuous from the quantum continuum (Heisenberg 1958; Planck 1931; von Neumann 2001). These entanglements (Francis 2012; Juan et al. 2013; Plenio 2007), or blemishes in the quantum continuum, may be observed as shape, form, cause, effect, past, present, future, the passage of time, and thus reality; the result of a decoupling of quanta from the quantum (coherent) continuum which leaks out and then couples together in a knot of activity which is observed as a wave form collapse.

As based on the Copenhagen theory of quantum mechanics (Bohr, 1958, 1963; Heisenberg 1955, 1958), time and reality are a manifestation of wave functions and alterations in patterns of activity within the quantum continuum which are perceived by consciousness as discontinuous. Wave form collapse is always a matter of probability, and is non-local, indeterministic and a consequence of conscious observation, measurement, and entanglement. Consciousness and the act of measurement, therefore, are entangled with the quantum continuum and can after the continuum and the space-time manifold. Just as a Time Traveler can accelerate or slow down, contracting or dilating time and time-space, consciousness can accelerate or slow down, and time will slow down or speed up accordingly (Joseph 1996, 2000).

Heisenberg (1958), cautioned, however, that the observer is not the creator of reality: "Quantum theory does not introduce the mind of the physicist as part of the atomic event. But it starts from the division of the world into the object and the rest of the world. What we observe is not nature in itself but nature exposed to our method of questioning." Nevertheless, the act of knowing, of observing, or measuring, that is, interacting with the environment in any way, creates an entangled state and a knot in the quantum continuum described as a "collapse of the wave function;" a knot of energy that is a kind of blemish in the continuum of the quantum field. This quantum knot bunches up at the point of observation, at the assigned value of measurement and can be entangled.

The same principles would also apply to time travel. The act of moving through time would effect time and all local and even more distant events. Traveling through the past or the future would effect every moment of that future; however, exactly what those changes may be, are indeterministic and can only be described by a probability function.

In the Copenhagen model, objects are viewed as quantum mechanical systems which are best described by the wave function and the probability function. "The reduction of wave packets occurs when the transition is completed from the possible to the actual" (Heisenberg, 1958).

The measuring apparatus and the observer also have a wave function and therefore interact with what is being measured. The effect of this is obvious when its a macro-structure measuring a micro-structure vs a macro-structure measuring a macro-structure. According to the Copenhagen interpretation (Bohr, 1949, 1963; Heisenberg, 1958), it is the act of measurement which collapses the wave function. It is also the measurement and observation of one event which triggers the instantaneous alteration in behavior of another object at faster than light speeds; i.e. entanglement. However, that which takes place faster than the speed of light means that an event in the future can travel from the future into the present and then into the past. Therefore, the future can effect the present and the past.

Time and Quantum Physics

In contrast to Newton and Einstein, quantum mechanics concerns itself with the dynamical change of state and its probability coupled with the Schrödinger (1926) time equations which are both time dependent and time independent for particles and waves. The state-function specifies the state of any physical system as a specific time t. The Schrödinger time equations relates states at a series time t^1 to a later time t^2.

In quantum mechanics, the Schrödinger (1926) equation is a partial differential equation that describes how the quantum state of a physical system changes with time. Like Newton's second law (F = ma), the Schrödinger equation describes time in a way that is not compatible with relativistic theories, but which supports quantum mechanics and which can be easily mathematically transformed into Heisenberg's (1925) matrix mechanics, and Richard Feynman's (2011) path integral formulation.

Therefore, time, in quantum physics, is not necessarily relative or even a temporal sequence, and the same is true of future and past. As summed up by Heisenberg (1958), "in classical theory we assume future and past are separated by an infinitely short time interval which we may call the present moment. In the theory of relativity we have learned that the future and past are separated by a finite time interval the length of which depends on the distance from the observer..." and where the past always leads to the future. However, "when quantum theory is combined with relativity, it predicts time reversal;" i.e. the future can lead to the past.

Time reversal, in quantum physics, is possible only in space smaller than 10^{-13}cm; smaller than the radii of an atomic nucleus--a Planck length. "The phenomenon of time reversal...belongs to these smallest regions" (Heisenberg 1958)

As detailed in previous chapters, as a space-time traveler accelerates he undergoes length contraction. Upon reaching near light speed, he may have shrunk in size below that of a Planck Length, as which point he will blow a hole through the fabric of space time due to the tremendous gravitational forces and energy releases, whereas contraction and time continues in a negative direction. As specified by quantum physics, it is only upon reaching near light speed that one shrinks to a size smaller than a Planck length; and then time travels in reverse.

Light, Wave Functions, Electrons, Positrons and Entanglement

Planck published his quantum hypothesis in 1900; that energy could only be absorbed or emitted as discrete quanta. Einstein (1905a,b,c) accepted Planck's conclusions and argued that light consists of quanta traveling through space--and the energy of a quanta of light is equal to the frequency of light multiplied by Planck's constant. Thus light could be described as a wave or as energy quanta traveling through space--and this explained the well known phenomenon of diffraction and interference (due to a wave) and the photo-electric effect which had to be due to a particle (Einstein 1905a,b,c).

Light is not the same as time. Light carries information. Light is not the future or the past but a snapshot, or movie of the past, and like all movies, it can be edited, thereby becoming something different. Light also has a wave function and can act directly on matter even at a distance.

James Clark Maxwell found that in electric and magnetic forces and fields, which are now called electromagnetic waves, have a speed of 300,000 kilometers per second (186,000 miles), the speed of light waves. Maxwell's theories and equations incorporate these electrostatic physical "fields" to account for all electromagnetic interactions. Since mass can become energy and energy mass, the "field" is therefore a physical entity that contains energy and has momentum which can be transmitted across space; including empty space and to distant locations in space-time.

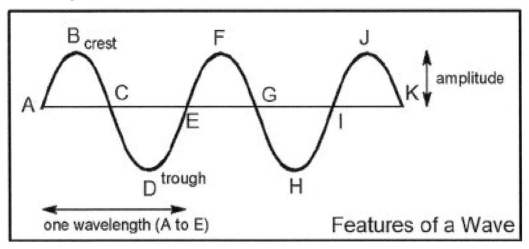

The electromagnetic spectrum, in order of increasing frequency and decreasing wavelength, consists of radio waves, microwaves, infrared radiation, visible light, ultraviolet radiation, X-rays and gamma rays.

Light is a duality, consisting of waves and particles (the photon or quanta). The energy of a photon is inversely proportional to its wavelength, such that longer wave lengths have little energy whereas short-wave lengths (such as X-rays) have a lot of energy. Light loses energy when moving away from the gravitational field of a star--causing a red shift.

Photons have a velocity which is the speed of light. Particles of light, photons (light quanta) have zero mass. Nevertheless, this mass-less photon has energy, as determined by it momentum which is the speed of light. A particle, even a massless particle has energy when it is in motion; i.e. kinetic energy. As demonstrated by Einsteins' famous equation, $E=mc^2$, energy can become mass.

Any kind of energy will contribute to the the mass; and this mass is equal to a given amount of energy divided by the square of the velocity of light. Therefore, every energy carries mass with it; and this implies that light, and time, have energy and potential mass.

Light is a form of electromagnetic radiation. Electromagnetic radiation (EMR) is a fundamental property of electromagnetism, propagating and traveling through space via photon wave particles, carrying radiant energy (Pollack & Stump, 2001; Slater & Frank, 2011). EMR has oscillating electric and magnetic fields which are entangled and have a fixed relationship to one another. EMR also acts on matter and is absorbed by matter and is emitted by charged particles.

EMR is characterized by the frequency or wavelength of its wave and has both momentum and angular momentum (Feynman 2011).

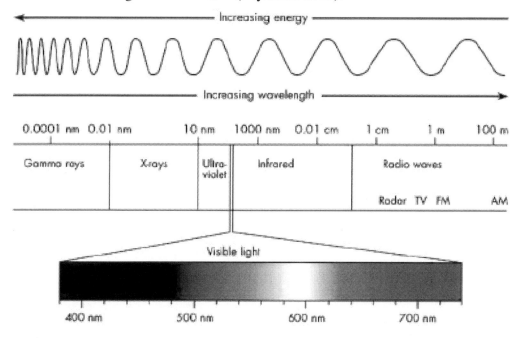

EMR exhibits a particle-wave duality (Dirac 1930, 1966a,b; Feynman 2011). As a wave, it is characterized by a velocity (the speed of light), wavelength, and frequency. When considered as particles, they are known as photons, The photon is the quantum of the electromagnetic interaction, and is the basic "unit" or constituent of all forms of EMR. High frequency (high energy) photons behave more like particles than lower-frequency photons do. The quantum nature of light is apparent at high frequencies (thus high photon energy).

In classical physics, EMR is considered to be produced when charged particles are accelerated by forces acting on them (Pollack & Stump, 2001; Slater & Frank, 2011). Since photons interact with single atoms and molecules, light can therefore act on and be acted on and can transfer energy to the matter which it acts on, even re-arranging single atoms and molecules, including doubly bonded macro-molecules such as DNA, depending on the amount of energy the photon carries. These effects include thermal heat and physical damage including the destruction of chemical bonds. As frequency increases beyond visible into the ultraviolet, photons have so much energy they can alter and injure doubly bonded molecules including DNA causing profound damage which can propagate through the organism.

The basic structure of matter involves charged particles bound together in many different ways (Feynman 1965, 2011). When electromagnetic radiation is incident on matter, it causes these charged particles to oscillate and gain energy.

Light and its particles (photons) interact with its environment.

At the higher end of the ultraviolet range, the energy of photons becomes large enough to impart enough energy to electrons to cause them to be liberated from the atom, in a process called photoionisation (Feynman 2011; Pollack & Stump, 2001; Slater & Frank, 2011).

Electrons are responsible for emission of most EMR and are responsible for producing much of the highest frequency electromagnetic radiation observed in nature. Electrons in turn can scatter photons (demonstrating light is quantized), whereas conversely electrons can be ejected and accelerated by photons (that is, the frequency not the intensity of light). Because electrons have low mass they are easily accelerated including, possibly to superluminal speeds thereby becoming positrons (Feynman 1949, 2011).

As proposed by Stueckelberg and later, Feynman, when electrons exceed the speed of light, they become positrons (Feynman 1949, 2011). Positrons are believed to move forward and backward in time. However, unlike tachyons, positrons have a positive charge. Wheeler seized upon the possible existence of positrons to explain why electrons appeared the same, suggesting that "they are all the same electron" with a complex, self-intersecting worldline (Feynman 1965, 2011). When they differ, they do so because they become positrons.

The first positron was discovered in 1932 by Carl Anderon for which he won the Nobel Prize for Physics in 1936. However, its theoretical history can be traced to Paul Dirac (1928) who proposed that electrons can have both a positive charge and negative energy. His proposals are now known as the Dirac equation which is a unification of relativity with quantum mechanics (Dirac 1930, 1966a; Feynman 2011); a unification which predicts time reversal (Heisenberg 1958) and the generation of negative energy and negative mass.

Whereas relativity and other branches of physics have attempted to ignore negative energy and negative mass, the duality between the positive and the negative and slower than light and faster that light particles is completely compatible with quantum mechanics (Dirac 1966a; Feynman 2011). Quantum formulations allow for electron and other particles to leap between positive and negative energy states and to journey at superluminal speeds.

Hence, light has a wave function and a particle function and interacts with and alters atoms and molecules, and can accelerate electrons which (theoretically) can assume a velocity beyond light speed thereby becoming a positron heading toward the past, such that the future can effect the past. Presumably, the same could be said of time. Time cannot be separate from the continuum except when perceived as such, thereby inducing a collapse of the wave function of time; experienced as the present, past, or future. Time is subject to entanglement, otherwise, entangled effects could not occur simultaneously or at faster than light speeds. "A" future can therefore effect "a" past and change it through entanglement and by effecting the wave function.

Relativity, Space Time...

Given entanglement, spooky action at a distance and the reality of the wave function, then the laws of physics must allow for information and effects to be conveyed faster than light speed and from the future to the past. If time is considered as a gestalt and a continuum and not a series of fragments, then the future and past are coextensive.

The quantum continuum is without dimensions and encompasses space and time in its basic unity of oneness. Everything within the quantum continuum can be effected by local effect and distant effects simultaneously at and beyond light speeds. Therefore, the future, and the "present" being part of this continuum can effect the past by effecting the wave function of the past, present, future, and thus, the space-time continuum, as all are entangled.

Probabilities and The Wave Function of the Time Traveler

According to quantum mechanics the subatomic particles which make up reality, or the quantum state, do not really exist, except as probabilities (Born et al. 1925; Dirac 1966a,b; Heisenberg 1925, 1927). These "subatomic" particles have probable existences and display tendencies to assume certain patterns of activity that we perceive as shape and form. Yet, they may also begin to display a different pattern of activity such that being can become nonbeing and thus something else altogether.

The conception of a deterministic reality is rejected and subjugated to mathematical probabilities and potentiality which is relative to the mind of a knower which registers that reality as it unfolds, evolves, and is observed (Bohr 1958, 1963; Heisenberg 1927, 1958). That is, by measuring, observing, and the mental act of perceiving a non-localized unit of structural information, injects that mental event into the quantum state of the universe, causing "the collapse of the wave function" and creating a bunching up, a tangle and discontinuous knot in the continuity of the quantum state.

Therefore, quantum mechanics, as devised by Niels Bohr, Werner Heisenberg, Dirac, Born and others in the years 1924–1930, does not attempt to provide a description of an overall, objective reality, but instead is concerned with quanta, probabilities and the effects of an observer on what is being observed. The act of measurement causes what is being measured to assume one for many possible values and yields the probability of an object or particle to be moving at one speed or direction or to be in one position or location, vs many others. Thus, it could be said that the act of observation causes a wave function collapse, a discontinuity in the continuum which is interpreted as reality and cause and effect.

Specifically, quantum physics is in part based on the fact that matter appears to be a duality, and can be both a wave and a particle; that is, to have features of both, i.e. particle-like properties and wave-like properties (Niel Bohr's complementary principle). Therefore, every particle has a wave function which describes it and which can be used to calculate the probability that a particle will be in a certain location or in a specific state of motion, but not both.

Hence, statements about both the position and momentum of particles can only assign a probability that the position or momentum will have some numerical value and this means that all the properties of an object can not be known at the same time but can only be described or deduced by assigning probabilities (Heisenberg's uncertainty principle). Moreover, according to the uncertainty principle, it is not possible to restrict any analysis to position or moment without effecting the other, and this is because the very act of eliminating uncertainty about position maximizes uncertainty about momentum (Heisenberg 1927). Uncertainty implies entanglement. Likewise, eliminating uncertainty about momentum maximizes uncertainty about position. Instead, one must assign a probability distribution which assigns probabilities to all possible values of position and momentum.

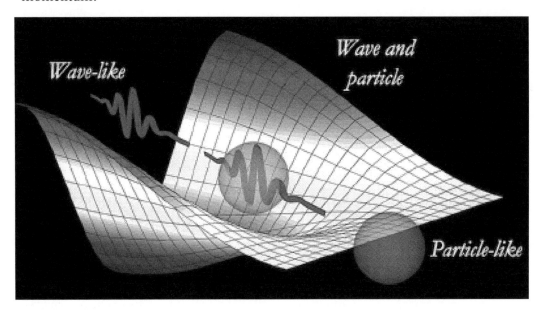

Therefore, no object, or particle, or quanta, or quantum, has its own eigenstate (inherent characteristic). Although every object appears to have a definite momentum, a definite position, and a definite time of occurrence, the object is in flux and it can't have a position and momentum at the same time. Therefore, when applied to time, then time, including the future and the past, can only be defined by a probability function and can only be related to events or objects in the past, present, or future in terms of the probability function. This means, the future and the past may change and that whatever is believed to have taken place or which will take place is best described in terms of probabilities.

Wave Functions

Only when the object can be assigned a specific value as to location, or time, or moment, does it have possess an eigenstate, i.e. an eigenstate for position, or an eigenstate for momentum, or an eigenstate for time; each of which is a function of the "reduction of the wave function;" also referred to as wave function collapse (Bohr, 1934, 1958, 1963; Heisenberg, 1930, 1955, 1958). Wave function collapse, which is indeterministic and non-local is a fundamental a priori principle of the Copenhagen school of quantum physics and so to is the postulate that the observer and the observed, and the past, present, and future, become entangled and effect one another.

Wave function collapse has also been described as "decoherence" which in turn leads to the "many-worlds" interpretation and the thought experiment known as "Schrödinger's Cat'" i.e. is a cat in a sealed box dead or alive? According to the Copenhagen interpretation, there is a 50% chance it will be dead and 50% chance it will be alive when it is observed, but one cannot know if it dead or alive until observed (measured). However, if there are two observers, one in the box with the cat the other outside the box, then the observer in the box knows if the cat is dead or alive, whereas the observer outside the box sees only a 50-50 probability (Heisenberg 1958).

The wave function describes all the various possible states of the particle. Rocks, trees, cats, dogs, humans, planets, stars, galaxies, the universe, the cosmos, past, present, future, as a collective, all have wave functions.

Waves can also be particles, thereby giving rise to a particle-wave duality and the the Uncertainty Principle. Particle-waves interact with other particle-waves. The wave function of a person sitting on their rocking chair would, within the immediate vicinity of the person and the chair, resemble a seething quantum cloud of frenzied quantum activity in the general shape of the body and rocking chair. This quantum cloud of activity gives shape and form to the man in his chair, and is part of the quantum continuum, a blemish in the continuum which is still part of the continuum and interacts with other knots of activity thus giving rise to cause and effect as well as violations of causality: "spooky action at a distance."

Likewise, the intrepid time traveler, journeying into the past, is also a wave function; particles and waves which interact locally with other local waves and creating additional blemishes in the quantum continuum. By traveling into the past or the future, the time traveler would come into contact with and change and alter the wave function of other blemishes in space-time. Hence, speeding into the past would therefore change the past, or rather, local events in that past, even if the Time Traveler sat still and did nothing at all except go with the flow. The wave function of the observer effects the wave function of what is observed and the wave function of immediate surroundings. The backward traveling time traveler effects each moment of "local" space-time as she travels through it. The wave function of the time traveler moving through time would spread out over space, becoming vanishingly small until disappearing.

By traveling into the past, the time traveler changes the past locally and perhaps even at a distance, depending on his actions. Likewise, since the future, present, and past are entangled, events taking place in the future, can effect and alter the past, thereby violating causality such that the past the time traveler visits may no longer be the past he was familiar with.

The probability function and entanglement when applied to the space-time continuum indicates that the, or rather "a" past may be continually changed and altered to varying degrees. This may also explain why memories of the past do not always correspond with the past record (Haber & Haber, 2000; Megreya & Burton 2008; Neisser & Harsch 1992). Although blamed on faulty memory, perhaps the past, and even memory of the past, have been and are continually and subtly being altered through entanglement and as predicted by the "many worlds" theory of quantum physics.

21: Paradoxes of Time Travel and the Multiple Worlds of Quantum Physics

Quantum mechanics, in theory, governs the behavior of all systems regardless of size (Bohr, 1934, 1947, 1958, 1963; Dirac 1966a,b; Heisenberg, 1930, 1955, 1958). Although it is said that macro-objects behave with certainty according to Newtonian physics, whereas micro-object don't, this is a function of relativity and the size of the measuring apparatus and the observer (Bohr, 1947, 1963; Heisenberg, 1927, 1930, 1958). For example, measuring systems and apparatus are macro- in size and which operate according to Newton Classical physics regardless of if what is being measure is micro or macro in size. A macro-size measuring device may not be sensitive to subtle changes in a macro-structure.

To truly and completely describe an event, object, or person, one must also determine the internal state, number of atoms, molecules, their arrangement, etc; all of which may not be measured when observing large structures as a whole (Heisenberg, 1958). Therefore, in regard to macro-structures it may be that if a macro-macro measuring device or observer were employed, it would be discovered that macro-scale macro-objects (and their component parts) do not behave with precise predictability, just as micro-scale objects when measured with a macro-measuring devices are subject to the Uncertainty Principle and can only be described in terms of probabilities. It may all depend on perspective and the proportions of what is doing the measuring/observing and what is being measured/observed. To truly measure a macro-object it must be measured from the micro-perspective and the macro-macro perspective (Heisenberg, 1958).

In a safe and comfortable finite universe proportion may not be a significant issue when measuring macro-objects using apparatus whose behavior is described by classical Newtonian physics. But from an infinite multi-dimensional "god's eye" view in a finite or infinite universe, all bets are off. Nevertheless, even with these caveats, most physicists would agree that Newton's laws are suitable for macro-objects whereas quantum mechanics must be applied for micro-structures if there is any hope of obtaining measurements and making predictions which might agree with experimental observation.

Central to quantum mechanics is the wave function (Bohr, 1963; Heisenberg, 1958). All of existence has a wave function, including light. Every aspect of existence can be described as sharing particle-like properties and wave-like properties. The wave function is the particle spread out over space and describes all the various possible states of the particle. According to quantum theory the probability of findings a particle in time or space is determined by the probability wave which obeys the Schrodinger equation. Everything is reduced to probabilities. Moreover, these particle/waves and these probabilities are entangled.

Reality is a manifestation of wave functions and alterations in patterns of activity within the quantum continuum which are entangled and perceived as discontinuous, and that includes the perception of time: past, present, future. The perception of a structural unit of information is not just perceived, but is inserted into the quantum state which causes the reduction of the wave-packet and the collapse of the wave function. It is this collapse which describes shape, form, length, width, and future and past events and locations within space-time (Bohr, 1963; Heisenberg, 1958).

In quantum physics, the wave function describes all possible states of the particle and larger objects, thereby giving rise to probabilities, and this leads to the "Many Worlds" interpretation of quantum mechanics (Dewitt, 1971; Everett 1956, 1957). That is, since there are numerous if not infinite probable outcomes, each outcome and probable outcome represents a different "world" with some worlds being more probable than others.

For example, an electron may collide with and bounce to the left of a proton on one trial, then to the right on the next, and then at a different angle on the third trial, and another angle on the fourth and so on, even though conditions are identical. This gives rise to the Uncertainty Principle and this is why the rules of quantum mechanics are indeterministic and based on probabilities. The state of a system one moment cannot determine what will happen next. Instead, we have probabilities which are based on the wave function. The wave function describes all the various possible states of the particle (Bohr, 1963; Heisenberg, 1958).

Since the universe, as a collective, must also have a wave function, then this universal wave function would describe all the possible states of the universe and thus all possible universes, which means there must be multiple universes which exist simultaneously as probabilities (Dewitt, 1971; Everett 1956, 1957). And the same would be true of time. Why shouldn't time have a wave function?

The wave function of time means there are infinite futures, presents, pasts, with some more probable than others.

Everett's Many Worlds

As theorized by Hugh Everett the universal wave function is "the fundamental entity, obeying at all times a deterministic wave equation" (Everett 1956). Thus, the wave function is real and is independent of observation or other mental postulates (Everett 1957), though it is still subject to quantum entanglement.

In Everett's formulation, a measuring apparatus MA and an object system OS form a composite system, each of which prior to measurement exists in well-defined (but time-dependent) states. Measurement is regarded as causing MA and OS to interact. After OS interacts with MA, it is no longer possible to describe either system as an independent state. According to Everett (1956, 1957), the only meaningful descriptions of each system are relative states: for example the relative state of OS given the state of MA or the relative state of MA given the state of OS. As theorized by Hugh Everett what the observer sees, and the state of the object, become correlated by the act of measurement or observation; they are entangled.

However, Everett reasoned that since the wave function appears to have collapsed when observed then there is no need to actually assume that it had collapsed. Wave function collapse is, according to Everett, redundant. Thus there is no need to incorporate wave function collapse in quantum mechanics and he removed it from his theory while maintaining the wave function, which includes the probability wave.

According to Everett (1956) a "collapsed" object state and an associated observer who has observed the same collapsed outcome have become correlated by the act of measurement or observation; that is, what the observer perceives and the state of the object become entangled. The subsequent evolution of each pair of relative subject–object states proceeds with complete indifference as to the presence or absence of the other elements, as if wave function collapse has occurred. However, instead of a wave function collapse, a choice is made among many possible choices, such that among all possible probable outcomes, the outcome that occurs becomes reality.

Everett argued that the experimental apparatus should be treated quantum mechanically, and coupled with the wave function and the probable nature of reality, this led to the "many worlds" interpretation (Dewitt, 1971). What is being measured and the measuring apparatus/observer are in two different states, i.e. different "worlds." Thus, when a measurement (observation) is made, the world branches out into a separate world for each possible outcome according to their probabilities of occurring. All probable outcomes exist regardless of how probable or improbable, and each outcome represent a "world." In each world, the measuring apparatus indicates which of the outcomes occurred, which probable world becomes reality for that observer; and this has the consequence that later observations are always consistent with the earlier observations (Dewitt, 1971; Everett 1956, 1957).

Predictions, therefore, are based on calculations of the probability that the observer will find themselves in one world or another. Once the observer enters the other world he is not aware of the other worlds which exist in parallel. Moreover, if he changes worlds, he will no longer be aware that the other world existed (Everett 1956, 1957): all observations become consistent, and that includes even memory of the past which existed in the other world.

The "many worlds" interpretation (as formulated by Bryce DeWitt and Hugh Everett), rejects the collapse of the wave function and instead embraces a universal wave function which represents an overall objective reality which consists of all possible futures and histories all of which are real and which exist as alternate realities or in multiple universes. What separates these many worlds is quantum decoherence and not a wave form collapse. Reality, the future, and the past, are viewed as having multiple branches, an infinite number of highways leading to infinite outcomes. Thus the world is both deterministic and non-deterministic (as represented by chaos or random radioactive decay) and there are innumerable futures and pasts.

As described by DeWitt and Graham (1973; Dewitt, 1971), "This reality, which is described jointly by the dynamical variables and the state vector, is not the reality we customarily think of, but is a reality composed of many worlds. By virtue of the temporal development of the dynamical variables the state vector decomposes naturally into orthogonal vectors, reflecting a continual splitting of the universe into a multitude of mutually unobservable but equally real worlds, in each of which every good measurement has yielded a definite result and in most of which the familiar statistical quantum laws hold."

DeWitt's many-worlds interpretation of Everett's work, posits that there may be a split in the combined observer–object system, the observation causing the splitting, and each split corresponding to the different or multiple possible outcomes of an observation. Each split is a separate branch or highway. A "world" refers to a single branch and includes the complete measurement history of an observer regarding that single branch, which is a world unto itself. However, every observation and interaction can cause a splitting or branching such that the combined observer–object's wave function changes into two or more non-interacting branches which may split into many "worlds" depending on which is more probable. The splitting of worlds can continue infinitely.

Since there are innumerable observation-like events which are constantly happening, there are an enormous number of simultaneously existing states, or worlds, all of which exist in parallel but which may become entangled; and this means, they can not be independent of each other and are relative to each other. This notion is fundamental to the concept of quantum computing.

Likewise, in Everett's formulation, these branches are not completely separate but are subject to quantum interference and entanglement such that they may merge instead of splitting apart thereby creating one reality.

Changing the Past: Paradoxes and the Principle of Consistency

Entanglement and "spooky action at a distance" prove that effects can occur faster than the speed of light (Lee et al. 2011; Matson 2012; Olaf et al. 2003), such that effects may take place simultaneously with or before the cause, such that the effect causes itself and may be responsible for the "cause;" a consequence of entanglement in the quantum continuum Likewise, a Time Travel can also effect the present and change the future of the past, or rather, "a" past or "a" future.

Since the time traveler and his time machine are comprised of energy and matter their presence and movement through time-space will also warp and depress the geometry of space-time thereby creating local and distant effects. Time travel would effect each local moment of time-space leading from one moment and location in time (e.g. the present) to another location, i.e. from the present to the future and from the future into the past, and these effects can occur simultaneously and at superluminal speeds.

Many physical systems are very sensitive to small changes which can lead to major change. Unless the past and the future are "hard wired" and already determined, then the very act of voyaging to distant locations in time will alter every local moment of that time continuum. In terms of "Many Worlds" the time traveler is continually creating or entering new worlds which exist in parallel. Each "world" becomes most probable the moment he interacts with the quantum continuum, including simply by passing through time.

As detailed by quantum mechanics (Dirac 1966a,b; Heisenberg, 1955), shape and form appear as blemishes and bundles of energy in the quantum continuum, the underlying quantum oneness of the cosmos, emerging out of the continuum but remaining part of it. According to the Copenhagen interpretation (Bohr 1934, 1963; Heisenberg, 1930, 1955), all quanta are entangled and therefore any jostling of one quanta can create an instantaneous ripple which can effect local as well as distant objects and events through intersecting wave functions.

The space-time continuum is part of that basic oneness and is the sum of its parts including what can and can't be observed. And this includes distant locations in space-time corresponding to all possible futures, presents, and pasts.

Quantum Physics of Time Travel

As pertaining to time travel, as the time traveler journeys through the quantum continuum of space-time, he will jostle and affect all the particles (or waves) he contacts as he passes through time, and these will effect particles and waves elsewhere in space-time, thus altering the very fabric of every local and perhaps more distant moments of space-time. In the "Many worlds" interpretation, the time traveler is not really changing the future of the past but is instead engaging in actions which cause branching and splitting, which leads him to a future and a past which exists in parallel with innumerable other futures and pasts. He is not changing the past, but entering a different past which always existed as a probability.

As predicted by the Many Worlds interpretation, if the Time Traveler did make a significant impact in the past, then the alteration of the past would effect the entire world-line of history related to that event, including the memories of everyone living since that event and all those who retain any knowledge of that event; such that no one would realize anything has changed.

Minds, consciousness, the brain, memory, are also part of the quantum continuum and can be altered by changes in it (Joseph 2010a). The act of observation can change an event and an event can alter the observing mind. If the past were changed, we would not know it had changed because everything related to that event would have changed, from the writing of books to documentary films about the event. The alteration of the quantum continuum is not limited to just that event but can alter the entire continuum, including the quantum composition of the brain and memories of everyone who has lived since that event (Everett 1956, 1957).

The Principle of Self-Consistency

Many theorists have argued that it is impossible to change the past. Igor Novikov and Kip Thorne (Friedman et al. 1990) called this the "self-consistency conjecture" and "the principle of self-consistency" and various paradoxes have since been proposed to support this contention such as: "what if you killed your grandmother before she gave birth to your mother? If you did, then you could not be born and could not go back in time to kill your grandmother! Presumably these paradoxes are supposed to prove it is impossible to travel back in time.

In some respects these "paradoxes" are the equivalent of asking: "What if you went into the past and grew wings?" And the answer is: "You can't." The time traveler can not go back in the past and grow wings, or an extra pair of hands, or develop super powers, and so on. Nor could the time traveler kill anyone in the past who, according to the past record, did not die on the date he was killed.

Just as in "real life" there are boundaries which prevent the average person from engaging in or making world-altering decisions, these same limitations would apply in the past. Therefore, according to the principle of self-consistency, it is impossible to change the past, and if any changes were made, they may be "local" rather than global, and thus completely non-significant and not the least memorable--just like daily life for 99.999999999% of the 7 billion souls who currently dwell on Earth and who live and die and are quickly forgotten except by a few other insignificant souls who are also quickly forgotten as if they never even exists. Any changes made in the past may be so insignificant as to be meaningless.

Just as it is impossible to determine position and momentum of a particle, the past may also be subject to imprecision such that by establishing certain facts, makes other facts less certain The past may also be subject to the Uncertainty Principle, which may explain why historians, eye-witnesses, and husbands and wives may not always agree about what exactly happened in the past or just moments before.

The Principle of Self-Consistency, however, holds that the past is hard wired and cannot be altered, and reverse causality is an impossibility (Friedman et al. 1990). By contrast, reverse causality (also referred to as backward causation and retro-causation) is based on the premise that an effect may occur before its cause, such that the future may effect the present and the present may effect the past. A "cause" by definition must precede the effect, otherwise the effect may negate the cause and the effect! For example, the if a man went back in time and killed his grandfather he would negate his own existence making it impossible to go back in time and kill his grandfather. On the other hand, if he did kill his "grandfather" it might turn out that his paternal lineage leads elsewhere, i.e. "grandmother" had an affair and another man fathered his own father. Thus killing his "grandfather" has no effect on his existence and does not interfere with his ability to go back in time to kill his grandfather. In this instance, the effect does not nullify the cause; which is in accordance with the principle of self-consistency. The past can't be altered and if it is, the result is not significant.

If the past is "fixed" and hard-wired and can't be altered, then although the time traveler may go back in time with the intention of killing his grandfather, or Hitler, or Lee Harvey Oswald, the result would be that he would be unable to do so; his gun would misfire, the bullet would miss, or he never got close enough to the intended victim to do the deed. The past is hard wired and can't be changed.

If the past can't be altered, then this also implies that the future may also be fixed and hard wired and is not subject to alteration. However, if the future is subject to change (as demonstrated by classical physics and the laws of cause and effect), then the future must exist in order to be altered; as predicted by quantum mechanics, entanglement, and Einstein's theories of relativity. If the future may be changed, then why not the past? According to the "Many worlds" interpretation, the past is not changed, instead one changes which past world becomes his reality.

The "Many Worlds" interpretation of quantum mechanics would allow one to kill their mother or commit a murder which had not taken place, in this "world;" but in so doing would be effecting the quantum continuum and contributing to the probability that an alternate world would become the time traveler's world once he commits these crimes.

Paradoxes and Many Worlds

Most time travel "paradoxes" are based on the premise that the time traveler some how gains powers or the will to do things he would never do, or to accomplish what others tried to do and failed. Even if the time traveler wanted to kill his mother before he was born, or assassinate Hitler before he came to power, would he be able to do it? Would he be able to get close enough to shove in that knife or fire that bullet? And if he did, maybe the victims would live. Maybe the knife or the bullet would miss the necessary organ. Maybe he would change his mind at the last moment. Maybe in the struggle someone else would shoot the Time Traveler in the head and he would die instead. Many people tried to kill Hitler and failed.

"Paradoxes" can be reduced to simple probabilities. What is the probability a time traveler would want to go back in time and kill his mother? What is the probability he would succeed? What is the probability others would intervene before he could do the deed? What is the probability that he might be killed in the attempt? ... and so on.

And if he did kill his mother, it would not be "his" mother.

An observer, object, particle, interacts with its environment, with the quantum continuum, changing and altering it. As postulated by "Many Worlds" theory, there is one ultimate reality, but many parallel realities and histories, like the branches of a tree, a hallway with infinite doors, or infinite highways all of which lead out of the city. One highway leads to a past where Hitler won the war. Another highway leads to a past where the Kennedy brothers were never killed. Yet another takes the time traveler to a world where he was never born.

According to quantum theory and the "many worlds" interpretation, a new highway, a new door, a new branch of the tree appears every time a particle whizzes by or an observer interacts with his environment, makes a decision, or records an observation. Thus, the time traveler may go back in time and kill the mother who dwells in a parallel world or universe, but he would be unable to kill his mother.

The "Many Worlds" Resolution of The Grandmother Paradox

Time Traveler "A" goes back into the past and kills his grandmother when she was still a little girl. An observer, object, particle, interacts with its environment, with the quantum continuum, changing and altering it. A time traveler going into the past would change every moment leading to that past simply by traveling through it, so that the past and the grandmother he encounters would be a different past and a different grandmother. As also predicted by the "Many Worlds" interpretation of quantum physics, a time traveler can appear in different parallel worlds. Therefore by killing this grandmother in this past time, Traveler "A" would be preventing the birth of that woman's time-traveling grandson "B", thereby preventing "B" from going into the past and killing the grandmother of the Time Traveler "A."

Multiple Paradoxes. Effects Negating Causes

A very wealthy scientist invents a time machine and travels 30 years back into the past to prevent the car accident which killed his very beautiful wife. He arrives in the parking lot of the business where she works and lets all the air out of her tires and disables the engine.

He visits the younger version of himself and gives him the blue print for building a time machine, and a list of 100 stocks and when to buy and sell them. The Time Traveler returns to the future.

When was the time machine invented?

His wife takes a cab to her Lover's apartment and that night they drive to her home and that of her husband (the younger version of the time traveler). The Lover discovers the blue print for the time machine and the list of 100 stocks.

Quantum Physics of Time Travel

The Lover and the wife sneak into the bedroom where her husband (the younger version of the time traveler) is napping and shove a knife through his heart.

Who invented the time machine?

The "Lover" upon killing the younger version of the Time Traveler (with the help of Time Traveler's wife), suddenly finds himself alone with the body, still holding the bloody knife in his hand. However, the Time Traveler's wife (and the blue print for a time machine and list of stocks) have disappeared. Upon his arrest he learns the Time Traveler's wife was killed hours before in a car accident.

Information Exists Before it is Discovered

Two research scientists, both bitter rivals, are competing to make a major scientific discovery. Scientist A, who is better funded, makes the discovery first, publishes the results, receives world wide acclaim and receives a Nobel Prize.

Scientist B loses all funding, does not get tenure, and is reduced to living in obscurity and working in his basement lab, where, 20 years later, he invents a time machine. Scientist B makes a copy of the article which won his rival, Scientist A, the Nobel Prize, and goes back in time and gives it to the younger version of himself. To ensure that the true inventor, Scientist A, does not get credit, the time traveler, Scientist B, kills Scienctist A.

The younger version of Scientist B publishes the discovery and receives all the credit and the Nobel prize. When the time traveler, Scientist B returns to his own time he is famous and has a Nobel prize on his shelf. When he looks at the scientific journal where the original article appeared, he sees the same article but with himself listed as the author. He no longer understands why he went back in time to kill his rival.

Who made the discovery?

Another scientist after laboring his entire life makes a major discovery which brings him wealth and world wide acclaim. However, he is old and sick and unhealthy and is unable to savor the honors, women, and riches which are now his for the asking but he is too old to enjoy. So, he invents a time machine, takes a copy of his notebook describing the discovery, goes back 50 years in time and gives it to his younger self, and explains: "here are the answers you are searching for. You are going to be rich and famous."

So where did the discovery come from?

Science is replete with examples of scientists who independently make the same discoveries although they were working independently of each other and often not knowing of the other's work (Merton, 1961; 1963; Hall, 1980). Examples include the 17th-century independent formulation of calculus by Isaac Newton, Gottfried Wilhelm Leibniz and others; the 18th-century discovery of oxygen by Carl Wilhelm Scheele, Joseph Priestley, Antoine Lavoisier and others; In 1989, Thomas R. Cech and Sidney Altman won the Nobel Prize in chemistry for their independent discovery of ribozymes; In 1993, groups led by Donald S. Bethune at IBM and Sumio Iijima at NEC independently discovered single-wall carbon nanotubes and methods to produce them using transition-metal catalysts. And the list goes on.

What this could imply, if the past and future are a continuum, is that the discovery exists before it is discovered, albeit in a distant location of space-time. Or, in terms of multiple worlds theory, one branch leads to a world where the discovery is made by scientist A, a different branch leads scientist B to the discovery. Yet another branch leads to a world where the discovery is not made until 20 years into the future, whereas a different branch leads to a world where it is discovered in just a few days.

Mozart heard his music in his head, already composed--and some have proposed there is a cosmic consciousness, a quantum continuum of consciousness, which contains all information, and that one need only a brain that can tap into select channels within this source to extract some of this information. If true, this may explain why discoveries are made simultaneously or why Mozart heard his music "already composed" in his head and then simply wrote it down.

Time Travel Through Many Worlds

As based on a Many Worlds interpretation of quantum physics, traveling backwards into the past would itself be a quantum event causing branching. Therefore the timeline accessed by the time traveller simply would be one timeline among many different branching pasts. Hence, the time traveler from one world/universe may kill his grandfather in another world/universe. Likewise, in the past of some worlds, Hitler won the war, the Kennedy brothers were never killed, the dinosaurs did not become extinct, mammals and humans never evolved, and so on. All quantum worlds, many worlds, all exist as there is an infinity of possible universes and worlds, each of which differs in some manner from the other, from the minute to the major.

However, by changing (or choosing) his past, the Time Traveler would not just be making this past "World" more probable, but may cause all pasts to become unified. That is, the other pasts disappear as they are subsumed by and merge to become this one unified past.

Therefore, according to the Many Worlds interpretation, by changing the past, and by creating a single unified past, then once the merging occurs, all "memories" of earlier branching events will be lost. No one will ever remember that there was any other past and no observer will even suspect that there are several branches of reality. As such, the past (and the future) becomes deterministic and irreversible, and this effects the wave function of time, such that the past shapes the future, and conversely, the future can shape the past.

Therefore, if a time traveler journeys to the past, his passage will either change the past so that those in the future can only remember the past that has been altered since this past is the past which leads up to them. Or, the past was never really altered and always included the Time Traveler's journey into the past. That is, this altered past has always existed even before he journeyed to it and this is because he traveled to and arrived in the past before he left. Thus everything he does, from the moment he left for the past, has already happened. The past, like the future is irreversible and has been hardwired into the fabric of space-time.

According to the Copenhagen model, one may predict probabilities for the occurrence of various events which are taking place or which will take place. In the many-worlds interpretation, all these events occur simultaneously. Therefore, the time traveler is not changing the past, but choosing one past among many: "new worlds" which always existed as probabilities.

And Yet Another Paradox: The Mirror World of the Past

There is yet another paradox which may plague the intrepid time traveler who journeys into the past. Because he has entered and has journeyed into a mirror world where time and space consists of negative energy and mass and where time runs in reverse, once he arrives in the past and exits his time machine, time may continue to run backwards. He may discover himself in a world where people go to sleep in the morning and awake at night, where Summer comes becomes Spring, and Spring comes becomes Winter, where Saturday is followed by Friday, where no one dies and the dead return to life and grow younger day by day, where World War II is followed by World War I, where the American Indians increase in number and White settlers retreat to the Eastern coasts, where the Spanish Conquistadors retreat to their ships and return to Spain, where the Great Pyramids are dismantled block by block, stone by stone, where mammals de-evolve into reptiles and reptiles into amphibians and life on land returns to the sea, and where each successive day takes him further into the long ago....

These are just some of the many conundrums which confront the time traveler.

22: Epilogue: A Journey Though The Many Worlds of Time

Half-drunk, I was bundled up against the frigid cold, about 100 yards downwind from the Abraxis Mars research station in Antarctica, near Scott's Base watching the southern lights, the aurora australis, playing across the frigid darkening skies. I took a gulp from the bottle of whiskey safely cradled in the thick glove of my right hand, and then another, welcoming the numbing of gradual oblivion.

My personal life was in tatters and I had been thinking about death; not suicide, but murder: killing my beautiful and brilliant wife. I had it all planned. It would be the perfect crime. Nobody would suspect a thing. Lifting the bottle I gazed down into the swirling liquid as if it held the secrets of life. She had betrayed me. Couldn't trust her. And I couldn't divorce her because she was a partner, half owner in Abraxis Thunder Bird Space Corporation. She knew all my secrets.

My beautiful brainy babe had me by the balls. I took another guzzle of whiskey.

Killing her would be easy. I had served in the military, as a Navy Seal. I had killed men with my bare hands.

But in my heart I knew it was all my fault. Was it any surprise she decided to get even? I fucked around, then she fucked around. Yeah yeah yeah, she apologized. Begged forgiveness. Still, my male pride couldn't handle it. Truth is, I still loved her.

That's why I left for the Abraxis Mars research station, to sort things out. Needed time to think. What better place than Antarctica? Now I was having second thoughts. It really was all my fault. My drinking, the carousing with anything that looked good in a short-skirt. Truth is: I needed her. She really was my best friend.

Its funny how you can love a woman so much and at the same time want to push her in front of a train. I took another slug of whiskey which burned soothingly and numbingly all the way down.

The auroras never ceased to fascinate me. The flowing cloud-like layers of billowing reds, greens, yellow, pinks, blues, purples, like an otherworldly alien invasion of color sweeping across the horizons.

I took another swallow of whiskey and that's when I saw the spiraling hoop-like rings of coiled light, like a giant searchlight from the heavens beaming down in a widening arc, crossing over the frozen tundra. It was coming closer, sweeping across the frozen wastes, and then, I was at the center of it, looking up through this narrowing black tunnel of swirling circling lights. I blinked in surprise: what I saw was me at the other end; like looking through the wrong end of a telescope. And then, like a bolt from the blue, I was lifted into the air then tossed like a rag doll onto the frozen ground, where I lay, the breath kicked out of me, dazed, confused, the world spinning, spinning. I lay there blinking as snow flakes lazily drifted down down down... and then I slowly sat up and looked to see if I was still in one piece. What the hell had happened? Standing up, brushing off the snow, I looked for the whiskey bottle. Where the hell was it? The wind was picking up and temperatures were dropping and I needed a drink, badly, so I began making my way back through the falling snow to the company research station, Abraxis One which was less than a 100 yards distant.

Abraxis One served one purpose: preparation for a human mission to Mars, and the frozen wastes of Antarctica was the most Mars-like place on Earth. It was my aerospace company, Abraxis, which was on the short-list of receiving the government contracts which would make a human mission to Mars a reality.

I entered the research station, closing the thick metal doors behind me which snapped shut with a loud metallic clang. As I brushed the snow from my heavy coat I sensed something was amiss. The interior of the research station looked different and several of my employees were surprised to see me, even though I had dinner with all of them the night before, surprising them all by my unexpected arrival. There were other clues that something was not right, like the arrangement of the furnishing, employees I had never seen before. There were paintings of what looked like a giant hub cap in space, instead of the panorama of possible Mars landing sites. And where were the pictures of the Martian mushrooms, lichens, and algae which proved there was enough water and a way to produce oxygen so to sustain life and a long term human mission to Mars?

"Somebody been redecorating?" I asked nobody in particular. And nobody answered. They just looked at me quizzically.

Well, I was still feeling dizzy and out of sorts from whatever happened out there on the tundra so I let it pass and headed to my room. Besides, I was still half-drunk.

The next morning, after I sobered up, I got my next surprise. All my luggage and gear were missing. All I had was the watch on my wrist and the clothes which I had stripped off before falling dead drunk into my bunk. I quickly dressed, rifled through my pockets and pulled out my wallet and a return plane ticket from Australia back to the USA. That was it! It was as if I had shown up without any luggage.

"What the hell?" I bellowed. "Where the fuck is my gear?" But nobody had an answer.

I paced around, shouting and grumbling, and then checked my watch but

Relativity, Space Time...

there was no signal. Finally my eyes fell upon a digital clock which was displaying the date.

"What the hell?" I shouted. "What's wrong with that clock?" But there was nothing wrong. I had lost two weeks. Swear to god I had only been here two days and somehow two weeks had gone by. My plane back to the USA would be leaving tomorrow morning!

Had I been hit by lightning out there on the tundra which addled my brain?

The next surprise. My corporate helicopter was not sitting outside on the helipad. I was fuming. "Where the fuck is my helicopter?" I demanded. But again, no one had an answer. They seemed as mystified by my presence, as I was by the missing luggage and helicopter. Everyone seemed surprised that I was even there.

No matter. We had more than one helicopter on the premises and I had one of the crew give me a lift to Williams Field where my corporate plane would be waiting to whisk me back to Australia and then the long flight back home. And then, another shock. When we arrived at Williams field, my helicopter was safely parked in its hangar but my corporate plane was no where to be seen. After a series of phone calls I was assured that the plane was sitting in the company hangar back at Melbourne, Australia.

"But that's impossible" I shouted. How the hell could I get from Australia to Williams Field in Antarctica, without a damn plane?"

"Somebody is fucking with me!" I growled.

I was near blinded with rage, demanded answers but got none. Again I checked my watch. No signal. Instead, I arranged to hitch a ride on a military transport and from there it was to the international airport in Melbourne where I would catch my flight to San Francisco.

Then the next surprise. Upon boarding the flight for home I discovered there was a little gray haired old lady sitting in my seat. I showed the stewardess my ticket. There was no mistake about that. Yet, the airline had no record of issuing it to me. Since the jet had several empty seats, that problem was soon solved. The stewardess brought me a whiskey, and I swallowed it in one gulp, thinking that maybe I should stop drinking. Maybe it was the alcohol creating problems with my memory? Or maybe I got hit with a lighting bolt when standing out there on the Antarctica tundra? I asked the stewardess for another whiskey, and then picked up an inflight magazine. That's when I got my next bolt from the blue.

Thumbing through I came upon a picture of Obama and the caption: "President Obama." I read the first couple of paragraphs and then stopped, and took a look at the cover of the magazine and the date. I turned to the fellow sitting next to me and showed him the article. "Is this a joke? When did Obama become president?"

My fellow passenger, with a face like a closed fist, gazed at me like he had a bad taste in his mouth. "He was elected in 2008."

"Right. But he's the vice president. Not the president."

My fellow passenger stared at me like I was nuts. "Biden is the vice-president. Obama is president."

This time I stared at him like he was crazy: "What are you talking about? Kennedy is president."

"Kennedy is dead" he replied giving me a once over with a single glance. "Obama is president now."

I suddenly felt sick. "Dead? When did this happen?"

He looked me up and down like I had just escaped from the loony farm and then muttered: "JFK was assassinated in 1964."

Suddenly I felt angry. "What the hell are you talking about? Kennedy died of old age in 2001. His son is president now."

My fellow passenger laughed uneasily and pointed at the empty whiskey glass I held in my hand. "You're confused, buddy. Maybe you out to lay off the booze. Both the Kennedy brothers were assassinated in the 1960s. Everybody knows that."

Now I was really angry. "What the hell are you talking about? Kennedy was reelected in 1968, then Nixon had two terms, and then JFK's brother Robert Kennedy was elected to two terms, and then Reagan got three terms, then there were a series of half-wits who warmed the president's chair for 16 years, and now Kennedy's son, JFK jr is president."

My fellow passenger stared at me like I was a bug in his coffee. Then he gathered up his bags and belongings and moved to another seat

Again I checked my watch. But damn, there was no signal. Instead, I picked up another magazine and then another scanning them quickly. Some of what I read was familiar. But so much more didn't make any sense. It was not what I remembered. It was as if the world had turned upside down and all the facts were rearranged. I felt dizzy. The world was spinning.

Had I lost my mind? Was I going senile?

Luckily I still had my watch. I gave it a voice command and waited for the holographic screen to emerge. I would get to the bottom of this pronto. Nothing happened. I spoke louder. Still no signal. Then I yelled at it. Nothing. Well, that's not quite true. Several of the other passengers were looking at me, and the stewardess was having a worried conversation with the fellow who had shared the seat beside me.

It was then that I began to suspect that I had crossed over into a parallel world, another universe. But how could this have happened? And then the answer came to me. It had to have been those spiraling rings of light out there on the frozen tundra. I remembered seeing a mirror image of myself at the far end of this black swirling tunnel of light, like staring into an Einstein-Rosen bridge leading to a mirror universe.

Relativity, Space Time...

Again I tried my watch. No hologram. No quantum library at my fingertips. Nothing. Then it dawned on me. Maybe they didn't have quantum computing in this world. I looked down at the magazines. Was this it? Magazine and newspapers? Were they still in the world of paper?

The stewardess, accompanied by the pilot came over to my seat. When I explained about my watch, the pilot gazed at it curiously and then asked me about "wearable technology" and when had this come out? Who made it, and so on. I quickly gathered, the people of this world were using "the internet;" a primitive form of technology which in my world had long ago gone the way of TV antennas.

After determining that my watch could not establish an internet connection, the stewardess graciously offered to let me borrow her "lap top" and I was soon online and in a state of shock. There could be no doubt. I had crossed over into a parallel world, a parallel universe with a history markedly different from my own. The two Kennedy brothers had been assassinated in the 1960s, JFK jr was killed in a plane accident in the 1990s. And there was no mission to Mars! The entire Kennedy space-program had been hijacked by corporate suits and greasy politicians who had shoved 100 billion dollars into the pockets of various corporations to build an International Space Station, that giant hubcap in the sky, whose main purpose, as near as I could tell, was to suck up money and spend tax payer dollars. And one of those companies sucking on that giant tit in the sky was Abraxis Thunder Bird Space Corporation. My company. But was it still my company?

There was a brief stop over at Honolulu International Airport in Hawaii, where I purchased several electronic devices. Then while we were winging our way to San Francisco International Airport I learned everything I could about "me."

Yes, I still owned Abraxis. Lived in the same expensive house, in the same exclusive neighborhood, on the same wide street. I checked the business news and did a search function on yours truly: There was the usual gossip about wine, women, and song. Ongoing lawsuits and contract disputes involving my company. More scandals. I read on and on. And then, as we were coming in for a landing at San Francisco International Airport, there it was: the obituary. My wife had died one week ago. I was stunned; like being hit in the stomach with a sledge hammer. She was dead! I quickly read the news about the "accident."

"The bastard killed her," I growled under my breath. It was murder, pure and simple. I know. I had planned the perfect crime. No one would ever suspect it wasn't an accident. But damn it, I had changed my mind. I loved that beautiful, brilliant bitch. I still loved her. And that bastard, my double who lives in this world, had killed her. Honestly, I had to choke back the tears. And then I began to think.

I rented a car at San Francisco International Airport and an hour later I was sitting in a BMW rental across the street from my palatial home in the Los Altos Hills. I parked so I could watch the heavy oak double front door of the three story house which looked like a small castle. I watched and waited then got out of the car which was shrouded in darkness. The lights in the house were still on. The bastard, my double, the "me" who lived in this world, was home. I waited, watched, kept thinking, making plans and then rejecting them. I wasn't sure what to do.

Was there any possibility I could go back to the world I came from? Not likely. Then what was I to do? How could I survive in this world?

As I thought about my options, a red sports car raced down the street, turned up the brick driveway and came to a screeching halt before the house. A girl got out. It was Jade, the daughter of one of my best friends. Eighteen years old and she looked good enough to eat. Yeah, she was a real dish and she came with forks and spoons. But I had never put my hands on her. Never even had a taste. Jade was a good girl. Smart. Bright future. I wasn't going to mess it up for her.

What the hell was Jade doing here at this hour?

The heavy oak double paneled front door opened and I could see "me" bathed in the porch light and framed by the lights inside the house. The "me," my double, was holding a bottle, probably whiskey. As Jade put her arms around him, he gazed out into the darkness, at her little red sports car, the brick driveway, out into the street, and then right at me. But I was so far away, and parked and standing in the shadows, I was sure he, my double, could not see me. Then he put his arm around Jade, drew her inside, and closed the heavy oak door.

That's when I decided to kill him. First off, the bastard was a murderer. He had killed my wife. I loved that brainy bitch. And he was fucking Jade, who I had known since she was 3 years old. Double bastard. Didn't matter she was 18. It wasn't right. Killed my wife. Fucking little Jade. Yes. He deserved to die. And, killing him would solve all my problems.

It was the smart thing to do for how else was I going to fit into this world? Start all over? Go to his door and ring the bell and say "hi" let me share your life. No. He was a killer. He had murdered my wife, and she my best friend and partner, the love of my life. There was only one answer: I would kill him and take his place.

Killing him would be easy. I had killed before. I knew his habits. His secret hideouts. The clubs where he liked to drink. The lonely mountain jogging paths he preferred. I had keys to the house, to his cars, private offices, and his private gym. And I had a way of disposing of the body: the incinerator in the basement labs of the headquarters of Abraxis which was not too far from Moffett Field and NASA Ames Research Center in Mountain View.

I did the deed the next morning as he stepped inside his offices; ice pick through the back of his neck and up into his brain. That night I burned the body.

Relativity, Space Time...

When I got home late that evening, I opened up a bottle of my best whiskey and was about to pour a drink when someone rang the front door bell.

"What the hell?" I muttered. I checked the video monitor. It was Jade. Beautiful sexy delicious Jade.

When I opened the heavy oak front door she was crying. Something about her boyfriend. She needed more advice. Someone to talk to.

"I told him what you said, but he just wouldn't listen! Then he hit me!" she cried.

Sobbing she through her arms around me and I reciprocated with my free hand, careful not to spill any of the whiskey. That's when I looked out into the yard, passed her red sports car, out into the street and there, in the darkened shadows I saw a car and a man standing next to it, looking back at me.

I gazed long and hard at him and then slowly closed the heavy oak front door.

I know who he is. Its me. Another double from an alternate universe. How many more will show up, I don't know. But what I do know: He is planning to kill me.

REFERENCES

Adler, R. J., Santiago, D. I., (1999) On Gravity and the Uncertainty Principle Mod.Phys. Lett. A14, 1371.

Aharony, O., et al., (2000). Large N Field Theories, String Theory and Gravity. Phys. Rept.323:183-386.

Aharonov, Y.; Komar, A.; Susskind, L. (1969). "Superluminal Behavior, Causality, and Instability". Phys. Rev. (American Physical Society) 182 ({5},): 1400–1403.

Aharonov Y., Albert D.Z., Vaidman L. (1988) How the Result of a Measurement of a Component of the Spin of a Spin-2 Particle Can Turn Out to be 100 Physical review letters 60, 1351.

Ahluwalia, D. V. (2000). Wave-particle duality at the Planck scale: freezing of neutrino oscillations. Physics Letters A 275, 31-35

Al-Khalili (2011). Black Holes, Wormholes and Time Machines, Taylor & Francis.

Almheiri, A. et al. (2013). Black Holes: Complementarity or Firewalls? J. High Energy Phys. 2, 062

Appelquist, T., Chodos, A. (1983). Quantum effects in Kaluza-Klein theories. Phys. Rev. Lett. 50, 141–145.

Aristotle, On Divination in Sleepung, C.G. (1989), Memories, Dreams and lections, New York: Vintage.

Arun,K. G., et al., (2006). Testing post-Newtonian theory with gravitational wave observations. lass.Quant.Grav.23:L37-L43.

Ashby, N. (2003). "Relativity in the Global Positioning System". Living Reviews in Relativity 6: 16.

Baibas, N. et al. (2005) Residence in mountainous compared with lowland areas in relation to total and coronary mortality. A study in rural Greece, J Epidemiol Community Health. 2005 April; 59(4): 274–278.

Bancaud, J.; Brunet-Bourgin, F.; Chauvel, P.; Halgren, E. (1994). "Anatomical origin of déjà vu and vivid 'memories' in human temporal lobe epilepsy". Brain : a journal of neurology 117 (1): 71–90.

Barker, J. (1967) Premonitions of the Aberfan Disaster", JSPR, 44, Issue: 734, Pages: 169-181

Becker, W. (2009) Neutron Stars and Pulsars, Springer, NY

Bekenstein, J. D., (1972).Black holes and the second law. Lettere Al Nuovo Cimento, 15, 737-740.

Bell, J. S. (1964) On the Einstein Podolsky Rosen Paradox, Physics 1, 3, 195–200 (1964)

Bell,J. S. (1966) On the problem of hidden variables in quantum mechanics, Rev. Mod. Phys. 38, 447

Bem, D. L. (2011). Feeling the future: Experimental evidence for anomalous retroactive influences on cognition and affect. Journal of Personality and Social Psychology, 100, 407-425.

Bertonea, B., Hooperb, D., Silk, J., (2005) Particle dark matter: evidence, candidates and constraints. Physics Reports, Volume 405 279-390.

Bethe, H. A., et al., (2003) Formation and Evolution of Black Holes in the Galaxy, World Scientific Publishing.

Bierman, D. J., & Scholte, H. S. (2002, August). Anomalous anticipatory brain activation preceding exposure of emotional and neutral pictures. Paper presented at the meeting of the Parapsychological Association, Paris, France.

Bilaniuk, O.-M. P.; Sudarshan, E. C. G. (1969). "Particles beyond the Light Barrier". Physics Today 22 (5): 43–51.

Blandford, R.D. (1999). "Origin and evolution of massive black holes in galactic nuclei". Galaxy Dynamics, proceedings of a conference held at Rutgers University, 8–12 Aug 1998,ASP Conference Series vol. 182.

Bo, L., Wen-Biao, L. (2010). Negative Temperature of Inner Horizon and Planck Absolute Entropy of a Kerr Newman Black Hole. Commun. Theor. Phys. 53, 83–86.

Bock, R. K. (9 April 1998). "Cherenkov Radiation". The Particle Detector BriefBook. CERN.

Bohr, N., (1913). "On the Constitution of Atoms and Molecules, Part I". Philosophical Magazine 26: 1–24.

Bohr, N., (1913). "On the Constitution of Atoms and Molecules, Part I". Philosophical Magazine 26: 1–24.

Bohr, N. (1934/1987), Atomic Theory and the Description of Nature, reprinted as The Philosophical Writings of Niels Bohr, Vol. I, Woodbridge: Ox Bow Press.

Bohr. N. (1949). "Discussions with Einstein on Epistemological Problems in Atomic Physics". In P. Schilpp. Albert Einstein: Philosopher-Scientist. Open Court.

Bohr, N. (1958/1987), Essays 1932-1957 on Atomic Physics and Human Knowledge, reprinted as The Philosophical Writings of Niels Bohr, Vol. II, Woodbridge: Ox Bow Press.

Bohr, N. (1963/1987), Essays 1958-1962 on Atomic Physics and Human Knowledge, reprinted as The Philosophical Writings of Niels Bohr, Vol. III, Woodbridge: Ox Bow Press.

Bokulich, A, & Jaeger, G. (2010) Philosophy of Quantum Information and Entanglement, Cambridge University Press.

Bonnor, W. Steadman, B.R (2005). "Exact solutions of the Einstein-Maxwell equations with closed timelike curves". Gen. Rel. Grav. 37 (11): 1833.

Born, M. Heisenberg, W. & Jordan, P. (1925) Zur Quantenmechanik II, Zeitschrift für Physik, 35, 557-615, 1925

Bottino, A., et al., (2003). Non-baryonic dark matter. Nuclear Physics B - Proceedings Supplements Volume 114, 27-37.

Boynton, R. (2001). "Precise Measurement of Mass". Sawe Paper No. 3147. Arlington, Texas: S.A.W.E., Inc

Bressi, G.; Carugno, G.; Onofrio, R.; Ruoso, G. (2002). "Measurement of the Casimir Force between Parallel Metallic Surfaces". Physical Review Letters 88 (4): 041804.

Brill, D. (2012) "Black Hole Horizons and How They Begin", Astronomical Review (2012).

Broughton, R. (1982) Human consciousness and sleep/waking rhythms: A review and some neuropsychological considerations. Journal of Clinical Neuropsychology, 4, 193-218.

Brougton, R. S. (1991). Parapsychology. Ballatine, New York.

Bruno, N. R., (2001). Deformed boost transformations that saturate at the Planck scale. Physics Letters B, 522, 133-138.

Buser, M. et al. (2013). Visualization of the Gödel universe. New Journal of Physics. Vol. 15.

Caldwell, N. et al., (2010) A Star In The M31 Giant Stream: The Highest Negative Stellar Velocity Known. The Astronomical Journal 139 372-377

Caldwell, R. Kamionkowski, M. (2009) Cosmology: Dark matter and dark energy. Nature 458, 587- 589.

Calvo, J. M., Badillo, S., Morales-Ramirez, M., Palacios-Salas, P. (1987) The role of the temporal lobe amygdala in ponto-geniculo- occipital activity and sleep organization in cats. Brain Research, 403, 22-30.

Campbell, J. (1988) Historical Atlas of World Mythology. New York, Harper & Row.

Carpenter, J. C. (2004). First sight: Part one, a model of psi and the mind. Journal of Parapsychology, 68, 217–254.

Carpenter, J. C. (2005). First sight: Part two, elaboration of a model of psi and the mind. Journal of Parapsychology, 69, 63–112.

Carroll, R. T. "The Skeptic's Dictionary"

Carroll, S (2004). Spacetime and Geometry. Addison Wesley.

Casagrande, V. A., & Joseph, R. (1978). Effects of monocular deprivation on geniculostriate connections in primates. Anatomical Records, 14, 2001.

Casagrande, V. A., & Joseph, R. (1980). Morphological effects of monocular deprivation and recovery on the dorsal lateral geniculate nucleus in Galago. Journal of Comparative Neurology, 194, 413-426.

Casimir, H. B. G. (1948). "On the attraction between two perfectly conducting plates". Proc. Kon. Nederland. Akad. Wetensch. B51: 793.

Chen, H.-W., Lanzetta, K. M., & Pascarelle, S. Spectroscopic identification of a galaxy at a probable redshift of $z = 6.68$. Nature, 398, 586 – 588 (1999).

Chen, S., and Jing, J. (2009). Strong field gravitational lensing in the deformed Hořava-Lifshitz black hole. Phys. Rev. D 80, 024036.

Cho, A. (2011) Furtive Approach Rolls Back the Limits of Quantum Uncertainty Science 333 690-3.

Chodos, A. (1985). "The Neutrino as a Tachyon". Physics Letters B 150 (6): 431.

Chou, C. W. et al. (2010). Optical Clocks and Relativity. Science, Vol. 329 no. 5999 pp. 1630-1633.

Chou, M-I. et al., (2009). A Two Micron All-Sky Survey View of the Sagittarius Dwarf Galaxy. http://arxiv.org/pdf/0911.4364.

Cooper, G. (2009). The Cell: A Molecular Approach. Washington, DC: ASM Press.

Daly, D. (1958) Ictal affect. American Journal of Psychiatry, 115, 97- 108.

Deffayet C., et al., (2002). Nonperturbative continuity in graviton mass versus perturbative discontinuity. Phys. Rev. D 65, 044-026.

DeWitt, B. S., (1971). The Many-Universes Interpretation of Quantum Mechanics, in B. D.'Espagnat (ed.), Foundations of Quantum Mechanics, New York: Academic Press. pp. 167–218.

DeWitt, B. S. and Graham, N., editors (1973). The Many-Worlds Interpretation of Quantum Mechanics. Princeton University Press, Princeton, New-Jersey.

Dietrich, M., et al., (2009) Black Hole Masses of Intermediate-Redshift Quasars: Near Infrared Spectroscopy. The Astrophysical Journal, 696, 1998-2013.

Dirac, P. (1928). "The Quantum Theory of the Electron". Proceedings of the Royal Society of London. Series A, 117 (778): 610–24.

Dirac, P. (1930) Principles of Quantum Mechanics

Dirac, P. (1966a) Lectures on Quantum Mechanics

Dirac, P. (1966b). Lectures on Quantum Field Theory .

Dyson, F. W. et al. (1920) Philos. Trans. R. Soc. London, Ser. A, 220, 291

Eadie, B. J., & Taylor, C. (1993) Embraced by the light. New York, Bantam

Economos AC, et al. (1982) Effects of simulated increased gravity on the rate of aging of rats: Implications for the rate of living theory. Arch Gerontol Geriatr. 1:349–363.

Einstein, A. (1905a). Does the Inertia of a Body Depend upon its Energy Content? Annalen der Physik 18, 639-641.

Einstein, A. (1905b). Concerning an Heuristic Point of View Toward the Emission and Transformation of Light. Annalen der Physik 17, 132-148.

Einstein, A. (1906a). On the Theory of Light Production and Light Absorption. Annalen der Physik 20, 199-206.

Einstein, A. (1906b). The Principle of Conservation of Motion of the Center of Gravity and the Inertia of Energy. Annalen der Physik 20, 627-633.

Einstein, A. (1907). On the Relativity Principle and the Conclusions Drawn from It. ahrbuch der Radioaktivität, 4, 411–462.

Einstein, A. (1910). The Principle of Relativity and Its Consequences in Modern Physics. Archives des sciences physiques et naturelles (ser. 4), 29, 5–28, 125–244.

Einstein, A. (1911). On the Influence of Gravitation on the Propagation of Light, Annalen der Physik (ser. 4), 35, 898–908,

Einstein, A. (1913). Physical Foundations of a Theory of Gravitation, Naturforschende Gesellschaft, Zürich, Vierteljahrsschrift, 58, 284–290

Einstein, A. (1914). On the Foundations of the Generalized Theory of Relativity and the Theory of Gravitation, Physikalische Zeitschrift, 15, 176–180.

Einstein, A. (1915a). Fundamental Ideas of the General Theory of Relativity and the Application of this Theory in Astronomy, Preussische Akademie der Wissenschaften, Sitzungsberichte, 1915 (part 1), 315.

Einstein, A. (1915b). On the General Theory of Relativity, Preussische Akademie der Wissenschaften, Sitzungsberichte, 1915 (part 2), 778–786, 799–801.

Einstein, A. (1926). Letter to Max Born. The Born-Einstein Letters (translated by Irene Born) Walker and Company, New York.

Einstein A. (1939) A. Einstein, Ann. Math. 40, 922.

Einstein (1955) In a letter to Michelangelo Besso, 21 March 1955.

Einstein, A. and Rosen, N. (1935). "The Particle Problem in the General Theory of Relativity". Physical Review 48: 73.

Einstein A, Podolsky B, Rosen N (1935). "Can Quantum-Mechanical Description of Physical Reality Be Considered Complete?". Phys. Rev. 47 (10): 777–780.

Einstein, A., Lorentz, H.A., Minkowski, H., and Weyl, H. (1923). Arnold Sommerfeld. ed. The Principle of Relativity. Dover Publications: Mineola, NY. pp. 38–49.

Einstein, A.1961), Relativity: The Special and the General Theory, New York: Three Rivers Press.

Eisberg, R., and Resnick. R. (1985). Quantum Physics of Atoms, Molecules, Solids, Nuclei, and Particles. Wily,

Eoin P. O'Reilly (2002). Quantum Theory of Solids (Master's Series in Physics and Astronomy) CRC Press.

Essen, L.; Parry, J. V. L. (1955). "An Atomic Standard of Frequency and Time Interval: A Cæsium Resonator". Nature 176 (4476): 280.

Everett, A., & Roman, T. (2012). TIme Travel and Warp Drives, University Chicago Press.

Everett , H (1956), Theory of the Universal Wavefunction", Thesis, Princeton University

Everett, H. (1957) Relative State Formulation of Quantum Mechanics, Reviews of Modern Physics vol 29, 454–462.

Ezzati, M. E. M. Horwitz, D. S. K. Thomas, A. B. Friedman, R. Roach, T. Clark, C. J. L. Murray, B. Honigman. (2011) Altitude, life expectancy and mortality from ischaemic heart disease, stroke, COPD and cancers: national population-based analysis of US counties. Journal of Epidemiology & Community Health, 2011; DOI: 10.1136/jech.2010.112938

Fabio Scardigli (1999) Generalized uncertainty principle in quantum gravity from micro-black hole gedanken experiment. Physics Letters B, Volume 452, Issues 1-2, 15 April 1999, Pages 39-44.

Feinberg, G. (1967). "Possibility of Faster-Than-Light Particles". Physical Review 159 (5): 1089–1105.

Feinberg, G. et al. (1961). "Law of Conservation of Muons". Physical Review Letters 6 (7): 381–383.

Feinberg, G. (1967). "Possibility of Faster-Than-Light Particles". Physical Review 159

(5): 1089–1105.

Fewster C J (2000) A general worldline quantum inequality" Class. Quant. Grav. 17 1897-1911

Feynman, R. (1949). "The Theory of Positrons". Physical Review 76 (76): 749.

Feynman, R. (1965). The Development of the Space-Time View of Quantum Electrodynamics (Speech). Nobel Lecture, Stockholm.

Feynman, R. (1994). The character of physical law. New York, NY: Modern Library.

Feynman, R P. (2011) The Feynman Lectures on Physics, Basic Books.

Finkelstein, D., (1958). "Past-Future Asymmetry of the Gravitational Field of a Point Particle". Phys. Rev. 110: 965–967.

Frank, M. G. (2012) Sleep and Brain Activity, Academic Press.

Friedman, J. et al. (1990). Cauchy problem in spacetimes with closed timelike curves". Physical Review D 42 (6): 1915.

Freud, S. (1900) The interpretation of dreams. In J. Strachey (Ed) Standard Edition (Vol 5. London: Hogarth Press.

Ford L H & Roman T A (1995) Averaged Energy Conditions and Quantum Inequalities" Phys. Rev. D 51 4277-4286.

Fouqué, P.; Solanes, J. M.; Sanchis, T.; Balkowski, C. (2001). "Structure, mass and distance of the Virgo cluster from a Tolman-Bondi model". Astronomy and Astrophysics 375: 770–780.

Frolkis VV, Muradian KK, Timchenko FN, Mozzhukhina TG. (1997) Effects of hypergravity stress on intensities of gaseous exchange, RNA and protein synthesis, thermoregulation, and survival of animals of different species. Sci Cosm Tech.3:16–21.

Fukuda, K. (2002) Sleep and Hypnosis, Vol 4(3), 111-114.

Fuller, Robert W. and Wheeler, John A. (1962). "Causality and Multiply-Connected Space-Time". Physical Review 128: 919.

Gamow, G. (1946) Rotating universe. Nature 158, 549.

Garay, L. J. (1995). Quantum gravity and minimum length Int.J.Mod.Phys. A10 (1995) 145-166

Gasser, J. ; Leutwyler, H. (1985). Chiral perturbation theory: expansions in the mass of the strange quark. Nucl. Phys. B; (Netherlands); Journal Volume: 250: 465-516.

Geiss, B., et al., (2010) The Effect of Stellar Collisions and Tidal Disruptions on Post-Main-Sequence Stars in the Galactic Nucleus. American Astronomical Society, AAS Meeting #215, #413.15; Bulletin of the American Astronomical Society, Vol. 41, p.252.

Ghez, A. M.; Salim, S.; Hornstein, S. D.; Tanner, A.; Lu, J. R.; Morris, M.; Becklin, E. E.; Duchene, G. (2005). "Stellar Orbits around the Galactic Center Black Hole". The Astrophysical Journal 620: 744.

Gibbons, G. W. (2002) "Cosmological evolution of the rolling tachyon," Phys. Lett. B 537, 1.

Giddings, S., et. al., (1994) Quantum Aspects of Gravity", Proc. APS Summer Study on Particle and Nuclear Astrophysics and Cosmology in the Next Millenium, Snowmass, Colorado, June 29 - July 14, 1994, arXiv:astro-ph/9412046v1.

Giddings, S. (1995). The Black Hole Information Paradox," Proc. PASCOS symposium/Johns Hopkins Workshop, Baltimore, MD, 22-25 March, 1995, arXiv:hep-th/9508151v1.

Gloor, P. (1992). Role of the amygdala in temporal lobe epilepsy. In J. P. Aggleton (Ed.). The amygdala. (pp. 505-538), New York: Wiley-Liss.

Gloor, P. (1997). The temporal lobe and limbic system. Oxford University Press, New York.

Gödel K (1949a) A remark about the relationship between relativity theory and idealistic philosophy Albert Einstein: Philosopher-Scientist (Library of Living Philosophers vol 7) ed P A Schilpp (Evanston, IL: Open Court) p 557

Gödel K (1949b) An Example of a New Type of Cosmological Solutions of Einstein's Field Equations of Gravitation, Rev. Mod. Phys. 21, 447

Gödel (1995) Lecture on rotating universes Kurt Gödel: Collected Works (Unpublished Essays and Lectures vol 3) ed S Feferman (Oxford: Oxford University Press)

Goldbach, C. (2006). Direct Detection of non-baryonic Dark Matter. In Mass Profiles and Shapes of Cosmological Structures. EAS Publications Series, 20, 209-216.

Gondolo, P. (2005). Non-Baryonic Dark Matter. In Alain Blanchard and Monique Signore (Eds). Frontiers of Cosmology. Springer. pp. 279-333.

Grant, W. B. (2010) An ecological study of cancer incidence and mortality rates in France with respect to latitude, an index for vitamin D production. Dermatoendocrinol. Apr-Dec; 2(2): 62–67.

Grebal, Eva K. (2004). The evolutionary history of Local Group irregular galaxies. in McWilliam, Andrew; Rauch, Michael (eds) Origin and evolution of the elements. Cambridge University Press. p. 234-254.

Griffiths, D. J. (2004), Introduction to Quantum Mechanics (2nd ed.), Prentice Hall,

Griffiths, D. J., (2008) Introduction to Elementary Particles, Wily.

Grinker, R. R., & Spiegel, J. P. (1945). Men Under Stress. McGraw-Hill.

Haber, R. N., Haber, L. (2000). Experiencing, remembering and reporting events. Psychology, Public Policy, and Law, 6(4): 1057-1097

Hafele, J. C.; Keating, R. E. (1972a). "Around-the-World Atomic Clocks: Predicted Relativistic Time Gains". Science 177 (4044): 166–168.

Hafele, J. C.; Keating, R. E. (July 14, 1972b). "Around-the-World Atomic Clocks: Observed Relativistic Time Gains". Science 177 (4044): 168–170.

Halgren, E. (1992). Emotional neurophysiology of the amygdala within the context of human cognition. In J. P. Aggleton (Ed.), The amygdala. (pp. 191-228), New York, Wiley-Liss.

Halkola, A., et al., (2006) Parametric Strong Gravitational Lensing Analysis of Abell 1689. Mon.Not.Roy.Astron.Soc.372:1425-1462.

Hall, R. A. (198). Philosophers at War, New York, Cambridge University Press, 1980.

Hamm PB, Billica RD, Johnson GS, Wear ML, Pool SL. 1998. Risk of cancer mortality among the Longitudinal Study of Astronaut Health (LSAH) participants. Aviation Space and Environmental Medicine 69(2):142–144.

Hamm PB, Nicogossian AE, Pool SL, Wear ML, Billica RD. 2000. Design and current status of the Longitudinal Study of Astronaut Health. Aviation Space and Environmental Medicine 71(6):564–570.

Haraldsson, E. (1975). Reported dream recall, precognitive dreams, and ESP. In Research in parapsychology 1974. Metuchen: Scarecrow Press.

Haraldsson, E. (1985). Representative national surveys of psychic phenomena. Journal of the Society for Psychical Research, 53, 145-158.

Halzen, F., Martin, AD (1985) Quarks and leptons: An introductory course in modern particle physics. American Journal of Physics. 53, 287.

Hansen, C. J. et al., (2004) Stellar Interiors - Physical Principles, Structure, and Evolution. Springer.

Hartle, J. B., (2003) Gravity: An Introduction to Einstein's General Relativity. Benjamin Cummings.

Hawking, S. W., (1988) Wormholes in spacetime. Phys. Rev. D 37, 904–910.

Hawking, S., (1990). A Brief History of Time: From the Big Bang to Black Holes. Bantam.

Hawking, S. (2005). "Information loss in black holes". Physical Review D 72: 084013.

Hawking, S. W. (2014). Information Preservation and Weather Forecasting for Black Holes. http://arxiv.org/abs/1401.5761

Heisenberg, W. (1925) Über quantentheoretische Umdeutung kinematischer und mechanischer Beziehungen, ("Quantum-Theoretical Re-interpretation of Kinematic and Mechanical Relations") Zeitschrift für Physik, 33, 879-893, 1925

Heisenberg, W. (1927), "Über den anschaulichen Inhalt der quantentheoretischen Kinematik und Mechanik", Zeitschrift für Physik 43 (3–4): 172–198,

Heisenberg. W. (1930), Physikalische Prinzipien der Quantentheorie (Leipzig: Hirzel). English translation The Physical Principles of Quantum Theory, University of Chicago Press.

Heisenberg, W. (1955). The Development of the Interpretation of the Quantum Theory, in W. Pauli (ed), Niels Bohr and the Development of Physics, 35, London: Pergamon pp. 12-29.

Heisenberg, W. (1958), Physics and Philosophy: The Revolution in Modern Science, London: Goerge Allen & Unwin.

Hildner, R., Brinks, D., Nieder, J. B., Cogdell, R. J., & van Hulst, N. F. (2013). Quantum Coherent Energy Transfer over Varying Pathways in Single Light-Harvesting Complexes. Science, 340(6139), 1448-1451.

Hobson, J. A, (2010). "REM sleep and dreaming: Towards a theory of protoconsciousness". Nature Reviews Neuroscience 10 (11): 803–13.

Hobson, J. A., Lydic, R., & Baghdoyan, H. A. (1986) Evolving concepts of sleep cycle generation: From brain centers to neuronal populations. Behavioral and Brain Sciences, 9, 371-448.

Hobson J.A., Pace-Schott E.F., Stickgold R. (2000). Dreaming and the brain: Toward a cognitive neuroscience of conscious states. Behavioral and Brain Sciences 23:793.

Hodoba, D. (1986) Paradoxic sleep facilitation by interictal epileptic activity of right temporal origin. Biological Psychiatry, 21, 1267-1278.

Houellebecq, M. (2001) The Elementary Particles, Vintage Books, NY.

Jaffe, R. (2005). "Casimir effect and the quantum vacuum". Physical Review D 72 (2): 021301.

Joseph, R. (1982). The Neuropsychology of Development. Hemispheric Laterality, Limbic Language, the the Origin of Thought. J. Clin. Psy. 44 4-33.

Joseph, R. (1986b). Confabulation and delusional denial: Frontal lobe and lateralized influences. J. Clin. Psy. 42, 845-860.

Joseph, R. (1988a) The Right Cerebral Hemisphere: Emotion, Music, Visual-Spatial Skills, Body Image, Dreams, and Awareness. J. Clin. Psy., 44, 630-673.

Joseph, R. (1988b). Dual mental functioning in a split-brain patient. J. Clin. Psy., 44, 770-779.

Joseph, R. (1992a) The Limbic System: Emotion, Laterality, and Unconscious Mind. The Psychoanalytic Review, 79, 405-456.

Joseph, R. (1992b). The Right Brain and the Unconscious. New York, Plenum.

Joseph, R. (1996). Neuropsychiatry, Neuropsychology, Clinical Neuroscience, 2nd Edition. Williams & Wilkins, Baltimore,

Joseph, R. (2010a) Quantum Physics and the Multiplicity of Mind: Split-Brains, Fragmented Minds, Dissociation, Quantum Consciousness. "The Universe and Consciousness", Edited by Sir Roger Penrose, FRS, Ph.D., & Stuart Hameroff, Ph.D. Science Publishers, Cambridge, MA.

Joseph, R (2010b) The Infinite Cosmos vs the Myth of the Big Bang: Red Shifts, Black Holes, and the Accelerating Universe. Journal of Cosmology, 6, 1548-1615.

Joseph, R. (2010c). The Infinite Universe: Black Holes, Dark Matter, Gravity, Acceleration, Life. Journal of Cosmology, 6, 854-874.

Joseph, R. (2011a) Dreams and Hallucinations: Lifting the Veil to Multiple Perceptual Realities "The Universe and Consciousness", Edited by Sir Roger Penrose, FRS, Ph.D., & Stuart Hameroff, Ph.D. Science Publishers, Cambridge, MA.

Joseph (2011b) Evolution of Paleolithic Cosmology and Spiritual Consciousness, and the Temporal and Frontal Lobes Journal of Cosmology, 2011, Vol. 14.

Juan Yin, et al. (2013). "Bounding the speed of `spooky action at a distance". Phys. Rev. Lett. 110, 260407.

Jung, C. G. (1945) On the nature of dreams. (Translated by R.F.C. Hull.), The collected works of C. G. Jung, (pp.473-507) Princeton: Princeton University Press.

Jung, C. G. (1964) Man and his symbols. New York. Doubleday.

Kaku, (1999). Introduction to Superstring and M-Theory (2nd ed.). New York, USA: Springer-Verlag.

Kellert, S. H. (1993). In the Wake of Chaos: Unpredictable Order in Dynamical Systems. University of Chicago Press.

Kerr, R P. (1963). "Gravitational Field of a Spinning Mass as an Example of Algebraically Special Metrics". Physical Review Letters 11 (5): 237–238.

King, C. (2014). Space, Time, and Consciousness. Cosmology, 16, 230-259.

Knecht, M. (2003). "The Anomalous Magnetic Moments of the Electron and the Muon". In B. Duplantier, V. Rivasseau. Poincaré Seminar 2002: Vacuum Energy – Renormalization. Progress in Mathematical Physics 30. Birkhäuser Verlag.

Krippner, S., Ullman, M., & Honorton, C. (1971). A precognitive dream study with a single subject. Journal of the American Society for Psychical Research, 65, 192-203.

Krippner, S., Honorton, C., & Ullman, M. (1972). A second precognitive dream study with Malcolm Bessent. Journal of the American Society for Psychical Research, 66, 269-279.

LaBerge, S. (1990). Lucid Dreaming: Psychophysiological Studies of Consciousness during REM Sleep". In Richard R. Bootzin, John F. Kihlstrom, Daniel L. Schacter (Eds.). Sleep and Cognition. Washington, D.C.: American Psychological Association. pp. 109–126

Lambrecht, A. (2002) The Casimir effect: a force from nothing, Physics World, September 2002

Lamon, W. L. (1911), Recollections of Abraham Lincoln, 1847-1885.

Lange, R., et al (2001) What Precognitive Dreams are Made of: The Nonlinear Dynamics of Tolerance of Ambiguity, Dream Recall, and Paranormal Belief, Dynamical Psychology, Vol. 6.

Langevin, P. (1911), "The evolution of space and time", Scientia X: 31–54

Lee, K.C., et al. (2011). "Entangling macroscopic diamonds at room temperature". Science 334 (6060): 1253–1256.

Libet B. (1989) The timing of a subjective experience Behavioral Brain Sciences 12 183-185.

Libet, B., Gleason, C. A., Wright, E. W., & Pearl, D. K. (1983). Time of conscious intention to act in relation to onset of cerebral activity (readiness-potential) the unconscious initiation of a freely voluntary act. Brain, 106(3), 623-642.

Lorentz, H. A. (1892), "The Relative Motion of the Earth and the Aether", Zittingsverlag Akad. V. Wet. 1: 74–79

Lossve A. & Novikov, I (1992). The Jinn of the Time Machine: Non-Trivial Self-Consistent Solutions. Classical and Quantum Gravity, 9, 2115.

Lu, E (2000). "Expedition 7 – Relativity". Ed's Musing from Space. NASA. http://spaceflight.nasa.gov/station/crew/exp7/luletters/lu_letter13.html

Peterson LE, Pepper LJ, Hamm PB, Gilbert SL. 1993. Longitudinal Study on Astronaut Health: Mortality in the years 1959-1991. Radiation Research 133:257–264.

Maccarone, T. J., et al., (2007). A black hole in a globular cluster Nature 445, 183-185.

Maggiore, M., (1993) A Generalized Uncertainty Principle in Quantum Gravity Phys. Lett. B304, 65-69.

Majewski, S. R. et al., (2003). A 2MASS All-Sky View of the Sagittarius Dwarf Galaxy: I. Morphology of the Sagittarius Core and Tidal Arms. Astrophys.J. 599 (2003) 1082-1115.

Martin, N. F.,., et al., (2004) A dwarf galaxy remnant in Canis Major: the fossil of an in-plane accretion onto the Milky Way. Mon.Not.Roy.Astron.Soc.348:12.

Masiero, A., et al., (2005), Neutralino dark matter detection in split supersymmetry scenarios. Nuclear Physics B, Volume 712, 86-114.

Matson, J. (2012) Quantum teleportation achieved over record distances, Nature, 13 August

Matthew. F. (2012). Quantum entanglement shows that reality can't be local, Ars Technica, 30 October 2012

Mazure, A. & Le Brun, V (2012). Matter, Dark Matter and Anti-matter. Springer. NY.

McClintock, J. E. (2004). Black hole. World Book Online Reference Center. World Book, Inc.

Mead, M. N. (2008),. Benefits of Sunlight: A Bright Spot for Human Health, Environ Health Perspect. 2008 April; 116(4): A160–A167.

Megreya, A. M., & Burton, A. M. (2008). Matching faces to photographs: Poor performance in eyewitness memory (without the memory). Journal of Experimental Psychology: Applied, 14(4): 364–372.

Melia, F. (2007). The Galactic Supermassive Black Hole. Princeton University Press. pp. 255–256.

Melia, F. (2003b). The Edge of Infinity. Supermassive Black Holes in the Universe. Cambridge U Press. ISBN 978-0-521-81405-8.

Merloni, A., and Heinz, S., (2008) A synthesis model for AGN evolution: supermassive black holes growth and feedback modes Monthly Notices of the Royal Astronomical Society, 388, 1011 - 1030.

Merton, R. K. (1961) "Singletons and Multiples in Scientific Discovery: a Chapter in the Sociology of Science," Proceedings of the American Philosophical Society, 105: 470–86.

Merton, R. K. (1963) Resistance to the Systematic Study of Multiple Discoveries in Science," European Journal of Sociology, 4:237–82, 1963.

Miller, M. C. & Hamilton, D. P. (2002) Production of intermediate-mass black holes in globular clusters. Mon. Not. R. Astron. Soc. 330, 232–240.

Minchin, R. et al. (2005). "A Dark Hydrogen Cloud in the Virgo Cluster". The Astrophysical Journal 622: L21–L24.

Minkowski, H. (1909), "Raum und Zeit", Physikalische Zeitschrift 10: 75–88.

Morris, M. S. and Thorne, K. S. (1988). "Wormholes in spacetime and their use for interstellar travel: A tool for teaching general relativity". American Journal of Physics 56 (5): 395–412.

Müller, H., Peters, A., Chu, S. (2010) Nature 463, 926 (2010).

Nairz, O. et al. (2003) "Quantum interference experiments with large molecules", American Journal of Physics, 71 (April 2003) 319-325.

NASA (2008), Current ISS tracking Data http://spaceflight.nasa.gov/realdata/tracking/index.html

National Physical Laboratory, (2007), "I feel 'lighter' when up a mountain but am I?", www.NPL.co.UK/ United Kingdom)

Neihardt, J. G. (1988) Black Elk Speaks. Nebraska, U. Nebraska Press.

Neisser, U., & Harsch, N. (1992). Phantom flashbulbs: False recollections of hearing the news about Challenger. In E. Winograd & U. Neisser (Eds.), Affect and accuracy in recall: Studies of "flashbulb" memories (Vol. 4, pp. 9–31). New York: Cambridge University Press.

Neumann, J. von, (1937/2001), "Quantum Mechanics of Infinite Systems. Institute for Advanced Study; John von Neumann Archive, Library of Congress, Washington, D.C.

Neumann, J. von, (1938), On Infinite Direct Products, Compositio Mathematica 6: 1-77.

Neumann, J. von, (1955), Mathematical Foundations of Quantum Mechanics, Princeton, NJ: Princeton University Press.

Nouicer, W. (2007). Quantum-corrected black hole thermodynamics to all orders in the Planck length. Physics Letters B, 646, 63-71.

Noyes, R., & Kletti, R. (1977). Depersonalization in response to life threatening danger. Comprehensive Psychiatry, 18, 375-384

Oguri, M. (2010). The Mass Distribution of SDSS J1004+4112 Revisited, Publ.Astron. Soc.Jap. 62 (2010) 1017-1024

O'Leary, R. M., O'Shaughnessy, R., Rasio,F.A. (2007). Dynamical interactions and the black-hole merger rate of the Universe Phys. Rev. D 76, 061504(R).

Ohanian, H. C., & Ruffini, R. (2013) Gravitation and Spacetime, Cambridge University Press.

O'Neill, B. (2014) The Geometry of Kerr Black Holes, Dover

Olaf, N.. et al. (2003) "Quantum interference experiments with large molecules", American Journal of Physics, 71 (April 2003) 319-325.

Oyama J. (1982) Metabolic effects of hypergravity on experimental animals. In: Miquel J, Economos AC, editors. Space Gerontology; NASA conference publication; 1982. pp. 37–51.

Pagel J. F. (2014) Dream Science: Exploring the Forms of Consciousness, Academic Press.

Palmer, J. (1979). A community mail survey of psychic experiences. Journal of the American Society for Psychical Research, 73, 221-251.

Parker, L. & Toms, D. (2009), Quantum Field Theory in Curved Spacetime: Quantized Fields and Gravity, Cambridge University Press.

Penfield, W. (1952) Memory Mechanisms. Archives of Neurology and Psychiatry, 67, 178-191.

Penfield, W., & Perot, P. (1963) The brains record of auditory and visual experience. Brain, 86, 595-695.

Penrose, R. (1969) Rivista del Nuovo Cimento.

Perlmutter, S. (2003) Supernovae, Dark Energy, and the Accelerating Universe. Physics Today 53 56, 53-62.

Perlmutter, S., Aldering, G., Della Valle, M., Deustua, S., Ellis, R. S., Fabbro, S., Fruchter, A., Goldhaber, G., Goobar, A., Groom, D. E. et al. (1998) Nature (London)391, 51-54.

Pitts GC, Bull LS, Oyama J. (1975) Regulation of body mass in rats exposed to chronic acceleration. Am J Physiol. 1975;228:714–717.

Planck, M. (1931). Consciousness. The Observer. January 25, 1931.

Planck, M. (1931). The Universe in the Light of Modern Physics.

Planck, M. (1932). Where Is Science Going?

Plenio, V. (2007). "An introduction to entanglement measures". Quant. Inf. Comp. 1: 1–51

Polchinski, J. (1998). String Theory, Cambridge University Press,

Pollack, G. L., & Stump, D. R. (2001), Electromagnetism, Addison-Wesley.

Potzel W. (1992) et al., Hyperfine Interact. 72, 195

Pound, R.V. Rebka, Jr. G.A. (1959) "Gravitational Red-Shift in Nuclear Resonance" Phys. Rev. Lett. 3 439–441

Pound, R. V. Rebka, G. A. (1960) Phys. Rev. Lett. 4, 337.

Preskill, J. (1994). Black holes and information: A crisis in quantum physics", Caltech Theory Seminar, 21 October. arXiv:hep-th/9209058v1.

Radin, D. I. (1997). Unconscious perception of future emotions: An ex- periment in presentiment. Journal of Scientific Exploration, 11, 163– 180.

Radin, D. I. (2006). Entangled minds: Extrasensory experiences in a quantum reality. New York, NY: Paraview Pocket Books.

Rawlins, M. (1978). Beyond Death's Door. Sheldon Press.

Reinhardt S. et al., (2007) Nat. Phys. 3, 861.

Renn, J. et al. (1997). "The Origin of Gravitational Lensing: A Postscript to Einstein's 1936 Science paper". Science 275 (5297): 184–6.

Rhine, L. E. (1954). Frequency and types of experience in spontaneous precognition. Journal of Parapsychology, 18, 93-123.

Rhine, L.E. (1969). "Case study review". Journal of Parapsychology 33: 228–66.

Rhine, L.E. (1977). Research methods with spontaneous cases. In B.B. Wolman (Ed) Handbook of Parapsychology (pp. 59-80). New York: Van Rostrand.

Rindler, W. (2001). Relativity: Special, General and Cosmological. Oxford: Oxford University Press.

Rindler W 2009 Gödel, Einstein, Mach, Gamow, and Lanczos: Gödel's remarkable excursion into cosmology Am. J. Phys. 77 498

Ring, K. (1980). Life at Death: A Scientific Investigation of the Near-Death Experience. New York: Quill.

Rodriguez, A. W.; Capasso, F.; Johnson, Steven G. (2011). "The Casimir effect in microstructured geometries". Nature Photonics 5 (4): 211–221.

Rogers, L. W. (1923). Dreams and premonitions. Chicago: Theo Book Co.

Ross, C. A., & Joshi, S. (1992). Paranormal experiences in the general population. Journal of Nervous and Mental Disease, 180, 357-361.

Ruffini, R., and Wheeler, J. A. (1971). Introducing the black hole. Physics Today: 30–41.

Russell, D. M., and Fender, R. P. (2010). Powerful jets from accreting black holes: Evidence from the Optical and the infrared. In Black Holes and Galaxy Formation. Nova Science Publishers. Inc.

Rutherford E. (1914). Capture and loss of electrons by a particle. Nature, 92: 347-347.

Rutherford E. (1920). Nuclear constitution of atoms, Nature, 105: 500-501.

Ryback, D. (1988). "Dreams That Came True". New York: Bantam Doubleday Dell Publishing Group, 1988.

Sabom, M. (1982). Recollections of Death. New York: Harper & Row.

Saltmarsh, H. F. (1934). Report on cases of apparent precognition. Proceedings of the Society for Psychical Research, 42, 49-103.

Santos, E. (2010) Space–time curvature due to quantum vacuum fluctuations: An alternative to dark energy? Physics Letters A, 374, 709-712.

Scardigli. F., (1999) Generalized uncertainty principle in quantum gravity from micro-black hole gedanken experiment. Physics Letters B, Volume 452, Issues 1-2, 15 April 1999, Pages 39-44.

Schmeidler, G. R. (1988). Parapsychology and psychology: Matches and mismatches. Jefferson, NC: McFarland.

Schmidt, B. P., Suntzeff, N. B., Phillips, M. M., Schommer, R. A., Clocchiatti, A. Kirshner, R. P., Garnavich, P., Challis, P., Leibundgut, B., Spyromilio, J., et al. (1998) Astrophys. J.507, 46-63.

Shibata, M., and Nakamura, T. (1995) Evolution of three-dimensional gravitational waves: Harmonic slicing case. Phys. Rev. D 52, 5428–5444.

Schneider. P. (2006). Extragalactic Astronomy and Cosmology. Springer.

Schödel, R., et al., (2006). From the Center of the Milky Way to Nearby Low-Luminosity Galactic Nuclei. Journal of Physics: Conference Series, 54.

Schriever, F. (1987). A 30-year "experiment with time:" Evaluation of an individual case study of precognitive dreams. European Journal of Parapsychology, 7, 49-72.

Schrödinger, E. (1926). "An Undulatory Theory of the Mechanics of Atoms and Molecules". Physical Review 28 (6): 1049–1070. Bibcode:1926PhRv...28.1049S. doi:10.1103/PhysRev.28.1049.

Schrödinger E; Born, M. (1935). "Discussion of probability relations between separated systems". Mathematical Proceedings of the Cambridge Philosophical Society 31 (4): 555–563.

Schrödinger E; Dirac, P. A. M. (1936). "Probability relations between separated systems". Mathematical Proceedings of the Cambridge Philosophical Society 32 (3): 446–452.

Schurger A., Sitt J., Dehaene S. (2012) An accumulator model for spontaneous neural activity prior to self-initiated movement PNAS DOI: 10.1073.

Sen, A. (2002) Rolling tachyon," JHEP 0204, 048

Sheehan, D. P. (Ed.). (2006). Frontiers of time: Retrocausation— Experiment and theory. Melville, NY: American Institute of Physics.

Sigurdsson, S. & Hernquist, L. (1993) Primordial black holes in globular clusters. Nature 364, 423–425.

Slater, J. C. & Frank, N. H. (2011) Electromagnetism, Dover.

Smolin, L. (2002). Three Roads to Quantum Gravity. Basic Books.

Sondow, N. (1988). The decline of precognized events with the passage of time: Evidence from spontanous dreams. Journal of the American Society for Psychical Research, 82, 33-51.

Sonner, J. (2013). "Holographic Schwinger effect and the geometry of entanglement"Physical Review Letters 111 (21). doi:10.1103/PhysRevLett.111.211603.

Spottiswoode, S. J. P., & May, E. C. (2003). Skin conductance prestimulus response: Analyses, artifacts and a pilot study. Journal of Scientific Exploration, 17, 617–641.

Stevenson, I. (1961). An example illustrating the criteria and characteristics of precognitive dreams. Journal of the American Society for Psychical Research, 55, 98-103.

Stickgold, R. & Walker, M. P (2010). The Neuroscience of Sleep, Academic Press.

Stowell, M. S. (1995). Researching precognitive dreams: A review of past methods, emerging scientific paradigms, and future approaches. Journal of the American Society for Psychical Research, 89, 117-151.

Stowell, M. S. (1997a). Precognitive dreams: A phenomenological study - Part I: Methodology and sample cases. Journal of the American Society for Psychical Research, 91, 163-220.

Stowell, M. S. (1997b). Precognitive dreams: A phenomenological study - Part II: Discussion. Journal of the American Society for Psychical Research, 91, 255-304.

Suddendorf, T. & Corballis , M. C. (2007)The evolution of foresight: What is mental time travel, and is it unique to humans? Behavioral and Brain Sciences 30, 299-351

Taiminen, T.; Jääskeläinen, S. (2001). "Intense and recurrent déjà vu experiences related to amantadine and phenylpropanolamine in a healthy male". Journal of Clinical Neuroscience 8 (5): 460–462.

Taylor, E. F. & Wheeler, J. A. (2000) Exploring Black Holes, Addison Wesley.

Thakur, J. (1998). Physical nature of the event horizon. Jounral of Physics, 51, 693-698.

Thalbourne, M. A. (1984). Some correlates of belief in psychical phenomena: A partial replication of the Haraldsson findings. Parapsychological Review, 15, 13-15.

Thalbourne, M. A. (1994). The SPR centenary census: II. The survey of beliefs and experiences. Journal of the Society for Psychical Research, 59, 420-431.

Thorne, K. (1994) Black holes and time warps, W. W. Norton. NY.

Thorne, K. S. & Hawking, S. (1995). Black Holes and Time Warps: Einstein's Outrageous Legacy, W. W. Norton.

Thorne, K. et al. (1988). "Wormholes, Time Machines, and the Weak Energy Condition". Physical Review Letters 61 (13): 1446.

Twain, M. (1906) The Autobiography of Mark Twain, Volume 1, p. 274-276: transcript of 13 January 1906 dictation;

Ullman, M., Krippner, S., & Vaughan, A. (1977). Traumtelepathie (Dream telepathy 1973). Freiburg: Aurum.

Ullman, M., Krippner, S., & Vaughan, A. (1989). Dream telepathy: Experiments in nocturnal ESP (2nd ed.). Jefferson, NC US: McFarland & Co.

van der Wel, A.; et al (2013). "Discovery of a Quadruple Lens in CANDELS with a Record Lens Redshift". ApJ Letters. arXiv:1309.2826

Vessot R. F. C. et al., (1980) Phys. Rev. Lett. 45, 2081.

Vestergaard, M (2010). Black-hole masses of distant quasars. 2007 Spring Symposium on "Black Holes" at the Space Telescope Science Institute. Cambridge University Press.

Vestergaard, M., and Osmer, P. S., (2009). Mass Functions of the Active Black Holes in Distant Quasars from the Large Bright Quasar Survey, the Bright Quasar Survey, and the Color-Selected Sample of the SDSS Fall Equatorial Stripe. Astroph. Journal 699, 800-816.

Villard, Ray (2012). "The Milky Way Contains at Least 100 Billion Planets According to Survey". HubbleSite.org.

von Laue, Max (1913). "Das Relativitätsprinzip (The Principle of Relativity)". Jahrbücher der Philosophie 1: 99–128.

von Neumann, J. (1937/2001), "Quantum Mechanics of Infinite Systems. Institute for Advanced Study; John von Neumann Archive, Library of Congress, Washington, D.C.

von Neumann, J. (1938), On Infinite Direct Products, Compositio Mathematica 6: 1-77.

von Neumann, J. (1955), Mathematical Foundations of Quantum Mechanics, Princeton, NJ: Princeton University Press.

Warren, W. (1998) MR Imaging contrast enhancement based on intermolecular zero quantum coherences Science 281 247.

Wells, H. G. (1895) The Time Machine.

Wheeler, J. A.; Feynman, R. P. (1945). "Interaction with the Absorber as the Mechanism of Radiation". Reviews of Modern Physics 17 (2–3): 157–161.

Wheeler, J. A.; Feynman, R. P. (1949). "Classical Electrodynamics in Terms of Direct Interparticle Action". Reviews of Modern Physics 21 (3): 425–433.

Wheeler, J. A. (2010) Geons, Black Holes, and Quantum Foam. W. W. Norton.

Wilf, M., et al. (2007) The Guidebook to Membrane Desalination Technology: Reverse Osmosis, Balaban Publishers.

Williams, D. (1956). The structure of emotions reflected in epileptic experiences. Brain, 79, 29-67.

Wiseman. R. (2011) Paranormality, Macmillan

Wolfgang C. Winkelmayer, MD, ScD; Jun Liu, MD, MS; M. Alan Brookhart, PhD. (2009) Altitude and All-Cause Mortality in Incident Dialysis Patients. JAMA, 301(5):508-512.

Yin, J. et al. (2013). "Bounding the speed of `spooky action at a distance". Phys. Rev. Lett. 110, 260407.

Young, K. (June 6, 2006). "The Andromeda galaxy hosts a trillion stars". New Scientist.

Zhang, J. (2010). Rediscussion on Black Hole Angular Momentum. International Journal of Theoretical Physics, 49, 224-231.

Printed in Great Britain
by Amazon